Basic Steps in *Astronomy*

To Barbara

Basic Steps in *Astronomy*

John Boulton

BLANDFORD PRESS
Poole Dorset

First published in the U.K. in 1979

Link House, West Street,
Poole, Dorset BH15 1LL

ISBN 0 7137 1012 8

British Library Cataloguing in Publication Data

Boulton, John
Basic steps in astronomy.
1. Astronomy
I. Title
520 QB43.2

ISBN 0-7137-1012-8

Set in 12/13pt V.I.P. Baskerville, printed and bound in Great Britain by
Fakenham Press Limited,
Fakenham, Norfolk

Contents

Introduction

Basic Steps in Astronomy is designed as an introductory level reading in the science of astronomy, both theoretical and observational. Mathematics have been kept to a minimum though some theoretical principles have been used to explain the reasoning behind the observational topics – the main theme of the book. The intention is to induce the beginner to carry out some simple experiments which will help to explain the modern concept of astronomy.

The first half of the book begins with a history of astronomy and the development of the astronomer's tools, expanding into a discussion of the astronomical telescope and some of the auxiliary equipment. The celestial co-ordinates are covered in this part and a brief section on the measurement of time is included. To conclude this, which is basically a theoretical background, is a chapter devoted to helping the beginner make useful observations. This blends into the second half of the book within which are details of some of the observing topics and objects that can be observed with a binocular or small telescope. We look firstly at the solar system, the Sun and planets. Next comes a discussion of the galaxies in a broad context, moving on to look at the stars in the galaxies in more detail. Lastly an appendix section lists some of the objects that the reader may wish to have a look at with a small optical instrument.

Because the book is not dealing with a specific topic of astronomy, but has within its pages a coverage of many different branches of observational astronomy in combination with some theory behind our present knowledge, it will be of value to all who have an interest in the subject.

JOHN P. BOULTON

1979

I

Basic History of Astronomy

Astronomy is a subject which receives a never ending influx of information. Wood is always being piled on the fire of man's burning urge to know more. Not a week passes but some new piece of information is added to the infinite jigsaw puzzle of the universe – a new satellite in orbit about one of the planets, or the recent discoveries of the ring systems around the planets Uranus and Jupiter. With these discoveries come the explanations and theories, which provide a never ending source of reading for both the amateur and the professional astronomer.

This book is however, dedicated to the people who help to furnish the fire with fuel, the observers who brave the crisp winter night air to spend hours out of doors with either a telescope or a binocular.

With time so very limited on large telescopes, many projects are forgotten by the professional – projects such as large scale variable star and double star observations, all night meteor watches, fireball patrols and many other topics. To make a good job of these observations requires such a large number of instruments and observers that the professional institutions could not hope to carry them all out without some help from the amateur. In the following pages the author hopes to stimulate the interest of the reader to the extent that he will also want to become part of one of the many amateur groups spread across the country, making regular contributions to our knowledge.

Astronomy, or star gazing as it would then have been, began when man's first priorities – food and survival – were relieved by the use of tools, and a shelter became a permanent cave. A communal cave dwelling would make life much easier. Food would be plentiful and defence simple where a group of, say, a hundred men were living in a small area. With increasing colony size came a need to control the people to a fine degree. The religion of the celestial objects was a medium by which the chieftains would control the masses. It was a natural medium and it held everyone in the same state of awe, of wonder at the marvels of Heaven. It would be natural to assume that there must be some Godly person controlling the Sun, Moon and Stars.

By this time, 5000–6000 years B.C. observations of the celestial objects revealed irregularities in the smooth pattern of the sky. Eclipses of the Sun and of the Moon were observed, as were variations in the pattern of the stars, some of the stars moving in a different manner from the rest. These were, of course, the planets. It eventually became vital to the well being of the leaders that these unusual phenomena were forecast. Many irregularities from the norm were thought upon as bad omens, or messages from the Gods controlling the sky objects. To predict the occurrence of such phenomena, it is vital that a thorough knowledge of the cycles and periods of the heavenly bodies is at hand. This can only be gained through observation over a long period of time. Observatories were built with the prime function of being able to tell the time accurately between the various phenomena, the time of the year, etc.

Possibly one of the best known observatories is Stonehenge near the small town of Amesbury in Wiltshire, England. Stonehenge is a circle of large stones with some weighing over 50 tons, erected in such a way that by standing at different places within the circle it is possible to tell the time of the year by noting the positions of the celestial objects. Many correlations have been found in the alignment of certain stones at Stonehenge. The rising and setting points of the Sun and Moon are examples. The best-known correlation is the alignment of one of the gaps between a trilithon and the heelstone over which the mid-summer sunrise takes place. Many monuments and constructions were built around this time, some of which bear no relation to astronomy. Some like the rows of stones near the town of Carnac in Brittany just consist of hundreds and sometimes thousands of stones arranged in rows with little or no relation to anything but their own reality. All of this was taking place up to about 1000 B.C. but by this time the Greek civilisation and others such as the Aborigines of Australia, the American Indians, the Aztecs, the Incas and peoples as far north as the

Greenland Ice Cap were recognising the patterns of the stars. Constellations were being developed into stories about the race's ancestry. The Greeks are particularly noted for this in their mythology. The story of the forefathers of the Greek civilisation from the beginning of the Earth and the Sky states that the Gods that ruled these also watched over the people, made the crops grow and placed figures of the country's heroes in the sky in the form of groups of stars and constellations. There were forty-eight original groups of stars incorporated into mythology and handed down with each generation of people. The Greek civilisation kept mainly to Gods, heroes and animals, while other races used more inanimate objects including types of stones and other everyday objects.

The period from about 1000 B.C. onwards can be termed the age of the philosophers. People were looking for more explanation for the wonders of the heavens, the whys, hows and whens. Around the fourth century B.C. Eudoxus suggested that the Earth must be at the centre of the heavens and that all other bodies will revolve around it. The original idea for this probably came from the earlier thought that the sky was a fixed dome over the Earth holding the stars and planets, the Sun and Moon in the proper place. The weakness of the theory was that if all the objects were fixed to the same sphere how were the different motions of the objects to be accounted for. Observations showed that each planet moved with a slightly different period and that the planets moved differently from the stars. Hiparchus tried to change the theory put forward by Eudoxus but could not find a suitable modification to account for the difference in the periods of rotation. Ptolemy, or more correctly Claudius Ptolemaeus, came up with a theory which fitted into the observations very well. He used as a basis for his theory the ideas of both Eudoxus and Hiparchus to form what was to be the peak of this train of thought. The Earth stayed at the centre of the universe around which revolved nine transparent crystal spheres. On the first three spheres were situated the Moon, Mercury and Venus. The next sphere held the Sun in place. The 5th, 6th and 7th spheres held the planets Mars, Jupiter and Saturn. The 8th sphere contained all of the fixed stars. The 9th and last sphere, called the Prime Mover, controlled the motions of the other eight spheres. This was all right, it took care of the different rotation periods but a further observational anomaly caused this theory to break down.

The problem was that, at certain times during which the planet was visible, it was sometimes observed to move in a backwards or retrograde motion, not always staying at the same altitude, often moving higher or lower than its normal position. To include this in the theory, each planet was placed on a small sphere called the Epicycle, the Epicycle being placed on the major sphere. By rotation of the Epicycle any motion of the object could be incorporated. In fact the system was so flexible that if necessary it could even cope with a square orbit!

It was Ptolemy who wrote the Almagest, thirteen volumes of astronomy. The Almagest was Ptolemy's conclusion to the work which he had carried out. For over a thousand years the Almagest (*Al* is Greek for the, *Magest* is Greek for greatest) was hailed as the correct solution to the universe. In much the same way we, today, look upon the Theory of Relativity by Albert Einstein.

It was not until the early sixteenth century that revision of the heavens began to take place. In 1473 Nicolaus Copernicus was born. Copernicus studied mathematics and astronomy, during which time he compiled the work he had carried out in a book which he called the *De Revolutionibus Orbium Caelestium*. This contained the evidence necessary for a complete change of the system and involved a revival of the view held by two earlier astronomers, Aristachus in the fourth century B.C. and Seleucus in the second century B.C. The theories of both had been dismissed as ridiculous. They stated that the Sun, and not the Earth, was at the centre of the universe, the planets moving around it. Copernicus's work took over 30 years to complete. He was so put off by the non-acceptance by the general public that he would not publish it. His view was that only someone who had an intimate knowledge of the science would ever take more than a casual look at the implications before putting the book down again. Only he and a number of his students knew about the advance which could result from the publication. It was only after a great deal of persuasion by some of his pupils that he at last decided to go ahead with the book. Copernicus, now 70 years old, lay dying at his home when he received the first copy of it. He never had the chance to see the finished work. The Copernican system was nearer to what we know today than any other theory previously put forward. However, there was one serious fault. All of the planets were still located on the Ptolemaic crystal spheres.

Around 1560 a new astronomer came on to the scene. This was Tycho Brahe, born in 1546. He began his career at law school but it was not until the solar eclipse of 1560 that his interest was aroused to the extent that he gave up his position at school to study astronomy. Tycho undertook many tasks throughout his life to further our knowledge. Much of his work was in the production of tables of positions for the planets and stars. One of Tycho's main discoveries was the nova or new star. Many people actually saw the star but quite simply dismissed it, while Tycho made accurate observations of its position and brightness. The star became visible on 11 October 1572 and faded in brightness over the next 3 weeks. For this reason Tycho suspected that the universe was not what was originally thought – if stars could brighten and fade like that there must be a deeper explanation for many observations. Tycho developed his own theory of the visible universe. His theory suggested that the Earth was at the centre of the system, the Sun and Moon revolving around the Earth. The big difference in this theory was that Tycho thought that the other planets revolved not around the Earth but around the Sun.

During 1576 King Frederick II of Denmark started the construction of an observatory for Tycho to work in. The observatory was built on the island of Hven not far from Helsingor. It was during this period at the observatory that Tycho suggested his theory of the planetary system. After some years at Hven Observatory the people of his own society rejected him, both his pension and observatory were taken from him and he was thrown out of Denmark by Frederick II's successor. Tycho finally resettled near Prague where he started to rebuild his observatory to continue the work of making a star catalogue of all the brighter stars. It was here that Tycho invited a young astronomer to his observatory to help with the catalogue. The astronomer was Joannes Kepler. Kepler arrived at the observatory in 1600 but a year later Tycho Brahe died, leaving all the work to the young assistant. Using both his own observations and those of Tycho, Kepler completed the catalogue of the brighter stars and then started to work on his own theory of the structure of the solar system.

From observations Kepler found three laws which could be applied to any object in orbit around the Sun. These three laws are still in use today.

Law 1. The path of every planet in its orbit about the Sun describes an ellipse with the Sun at one focus.

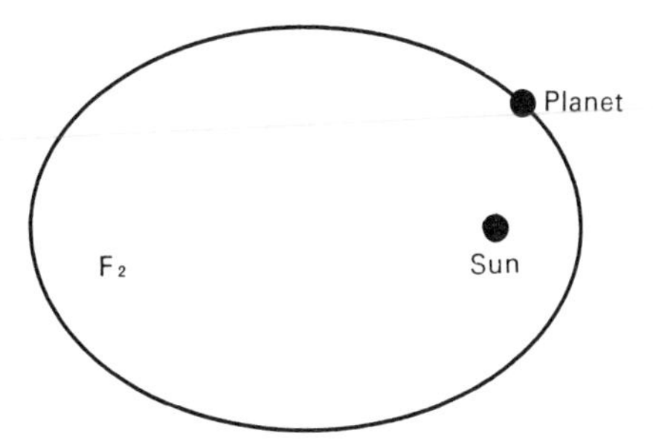

Fig. 1. Elliptical planetary orbit with the Sun at one focus and an empty F2.

Law 2. The speed of a planet varies with its distance from the Sun, so that a line drawn from the Sun to the planet sweeps out equal areas in equal periods of time.

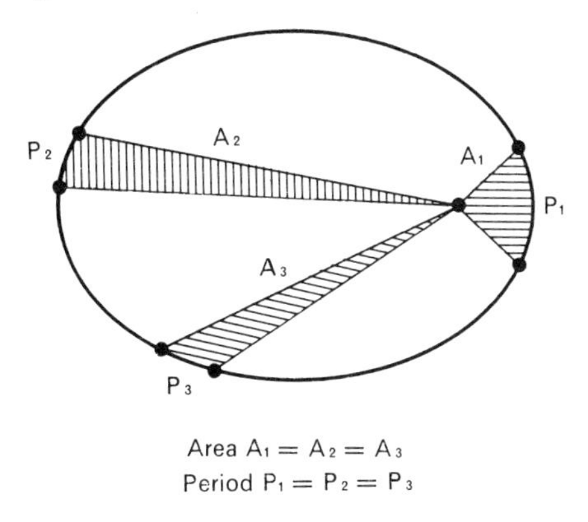

Fig. 2. The areas A_1, A_2 and A_3 are all equal and the time taken for the planet to traverse P_1, P_2 and P_3 are all equal, though the distances P_1, P_2 and P_3 are different.

Law 3. The period taken by a planet to orbit the Sun is proportional to the mean distance of the planet. The exact relation is, the period squared is proportional to the distance cubed.

These three laws changed the knowledge of the science of orbital mechanics to the extent that the planets could be made to travel around the Sun without anything holding them into place, such as the crystal spheres of Ptolemy. Kepler, however could not explain the force which held the planetary system together, but he certainly suspected that there was some mysterious force at work.

This is the end of the pre-telescopic period of astronomy because around 1610 the refracting telescope was invented. The first telescope to be used in astronomy was a simple arrangement of two spectacle lenses, one at each end of a length of organ pipe. It had a power of three magnifications. The man who was credited with this invention was of course Galileo Galilei. He used the telescope to observe the planets and stars, one of his findings was the three brightest moons of Jupiter, he later found the fourth which during his first observation was behind the planet and so was not visible. Galileo also observed the dark spots on the surface of the sun, they proved that the sun rotates and helped to further the Copernican idea of the Sun being at the centre of the solar system.

Galileo is best remembered for his experiments with mechanics. He noted that for a given length of a pendulum, no matter what the length of the stroke, the time between each stroke remained the same. He also proved that the force that causes objects to fall to the earth was the same for any mass of object. To prove this he dropped two heavy objects from the Tower of Pisa. One object was twice as heavy as the other but they both fell at the same rate. Galileo, developing the idea, realised that the reason that a feather and a stone fall at different rates is that the air resistance on the feather is much greater than on the stone, i.e. the feather has a much greater surface area to a given weight than the stone. Astronauts on the Moon tested this and found that on the airless surface, the stone and the feather do fall at the same rate.

This work and the observations which Galileo carried out reinforced the idea that all of the planets are in orbit around the Sun. For his persistence in this belief Galileo was summoned by the Inquisition to spend the rest of his life in seclusion. It was said that Galileo, after denouncing what he believed in, later turned to a friend and commented, 'Even so, it does move around the Sun'. In 1636 Galileo published the conclusion of his life study in *Dialogue of the New Sciences*.

During the time that Galileo was developing the telescope another man, Christiaan Huygens was also interested in this invention, from both the astronomical and the optical points of view.

Huygens was not born until after the invention and so could not claim any major part in the initial development but even today the Huygenian eyepiece is still in wide use. Huygens was an optician and spectacle maker by trade, this gave him the background to experiment with many types of eyepieces before arriving at the two plano-convex lenses for which he is now remembered.

All of the work which Galileo had done laid the foundation for the work of one of the greatest scientists this world has known, Sir Isaac Newton. Newton started his education at the grammar school in Grantham, Lincolnshire from which he went to Trinity College, Cambridge to study mathematics. Newton's first discovery was very much based on work by Galileo, who proved that the force holding objects to its surface was the same for all objects. From this Newton calculated the force required to keep the Moon in orbit and found it to agree very nearly with its observed rotation. To carry out these calculations Newton developed the method of 'Fluxions', our modern Calculus. Later Newton found that the reason that his calculations for the force between the Earth and the Moon only nearly agreed, was that the Moon also has some force to contribute to the system. He was able to arrive at a value for this force and from his studies found four laws which even though they are so simple, no scientist has ever found an exception to them. The first three are Newton's Laws of Motion, while the fourth is the Law of Universal Gravitation.

Newton's Laws of Motion

Law 1. Every body continues in a state of rest or motion in a uniform straight line unless acted on by a force.

Law 2. Every change of motion is proportional to the force causing it and takes place in the direction of the force.

Law 3. The action and reaction of two bodies upon each other is equal and opposite.

Law of Universal Gravitation

Every particle of matter in the universe attracts every other particle with a force which is proportional to the product of their masses and inversely proportional to their distance apart.

These laws, and much more of Newton's work, would never have been made known were it not for a man named Edmund Halley. Halley suspected that a newly found comet (Halley's comet) was actually in an orbit around the Sun and may have been seen many times before. To test his assumption out he went to Isaac Newton for his advice. Newton confirmed that it was almost certain that the comet would be in orbit and replied to Halley, 'Bring me three accurate positions for the comet on successive nights and I will calculate the orbit'. Halley, astounded at this, demanded to see the proof which Newton duly forwarded. It was only the persistence of Halley in insisting that Newton shared his discovery that the proof was released to the world. The findings of Isaac Newton revolutionised the concepts of astronomy, not only in the theoretical field, but in the observational field as well. By experimenting with lenses and mirrors in his study, Newton discovered that the main drawback of the refracting telescope, chromatic aberration, could be overcome by using a curved mirror as the main element in a telescope. After calculating the correct shape of the mirror to be a parabolic section, he started to figure a mirror from a metal disc. The material used was speculum, an alloy of one part tin and four parts copper. This telescope had one big advantage over its refractor counterpart, the mirror only required one accurately ground surface, whereas the refractor may have four surfaces in a two element compound objective. The telescope also had its disadvantages. The speculum metal tarnished very quickly and needed re-figuring and polishing at regular intervals. The eyepiece at the opposite end of the telescope to the refractor often encoursed a difficult and uncomfortable observing position. In spite of these disadvantages the reflector was developed and used by astronomers. The greatest advantage of the reflector was to show later in the development of telescopes in general. As the size of the instrument increased it was found that for little gain in the aperture of the refractor the weight of the lens and consequently the mounting had to be increased enormously. The reflector however could be built much larger without the excessive weight and at a much lower cost per inch of aperture than the refractor. The maximum limit for a refractor is thought to be the 40 in. (102 cm.) diameter Yerkes telescope but there is the possibility that a 236 in. (600 cm.) reflector is at work in the Caucasus mountains. The limit for large reflectors was thought to have been reached during the construction of the 200 in. (508 cm.) telescope on Mount Palomar but someday it may become possible to cast and grind extremely large telescope mirrors in space. At present there are plans to launch a telescope into orbit around the Earth, with a diameter of 100 in. (254 cm.).

Although Newton laid the foundation for planetary mechanics it was the French astronomer and mathematician Laplace who used the theory to a great extent and applied it to all the major bodies in the Solar System in five volumes of the *Mechanique Celeste*. Although Laplace lived to be 78 years old, this project was his life's work. He died in 1827 after completing the application of Newton's Laws of Mechanics and Gravitation.

There were anomalies between the predicted and the observed, which the Galilean/Newtonian mechanics could not explain. This led to a new theory, developed during the early twentieth century by a man named Albert Einstein. The theory is so simple in form but so complicated in its fullest extent, that at the time of publication it was said that only four or five people could use the theory in its entirety. It is of course the Theory of Relativity. Basically the theory assumes that space and time are intimate. Instead of treating the universe as being three dimensional, relativity incorporates four dimensions X, Y, Z and the fourth dimension T or time. Relativity, as the name suggests is based on relative observation. To explain this more simply and practically we can take as an example a train which is moving along a track with an arbitrary velocity. For this example the train's velocity must not be causing any air turbulence, i.e. the train is in a vacuum. An observer in the train, opens a window and drops a stone. Because there is no air outside the train, the observer will see the stone fall in a vertical path towards the embankment. This is simple enough, but now consider another observer standing on the embankment. To this observer the stone falls in a parabolic path towards the embankment. Which of the two observers is correct? The answer to this question is that before the experiment takes place the reference co-ordinate system must be chosen. The correct result is the result obtained from the chosen reference body.

Relativity has been put to use in predicting many irregularities the unusual motion of Mercury, the bending of star light by the gravity of the Sun. During the past years the theory has been changed, but only minor modifications have been made to bring the theory into line with observations. At the present time the Special and General Theories of Relativity are the ultimate in our theoretical knowledge. The Unified Field Theory, also created by Einstein, which incorporates the effects of electricity and magnetism into relativity has not yet been expanded to its fullest extent. Although much work is being done on this, the full implications are still very vague. Through the mid twentieth century no drastic changes have been made to our concept of astronomy, and our knowledge is increasing ever more rapidly. Small changes and many intricate modifications have been added to make the overall picture fit into the observed pattern of the universe. The major problem encountered in the study of astronomy is the fact that every problem solved leaves two more in its place.

2

The Tools of the Astronomer

In the last chapter we looked at the history of astronomy and into this came the early development of the telescope. In this chapter the aim is to show why each of the different types of telescope has been developed and what is the function of each type of instrument. First there will be a brief discussion of the function of the telescope and secondly a detailed description of instrument types.

The main aim of the telescope is to increase the amount of detail or increase the brightness of an astronomical object and present this to the human eye in the form of useful information, i.e. as an image. The aim is not, as is generally thought, to magnify the image – magnification is a secondary feature of the instrument, which must be taken into account. Objects which are of observational value in astronomy are usually too faint to be seen with the naked eye, and a telescope must gather as much light as possible and focus it into the pupil. The lens of the human eye is approximately 0·15 in. (4 mm.) in diameter and so the area that actually collects the light is then 0·018 in.^{2} (12·5 mm.^{2}). A small telescope lens with a 50 mm. diameter objective will have an area of 1963 mm.^{2} – 157 times the area of the eye. This means that the 50 mm. telescope will collect 157 times more light than will the human eye.

The first telescope to be used in astromony was the refractor. It consisted of two spectacle lenses, one fastened to each end of a length of organ pipe. Figure 3 shows the general configuration of this type of telescope.

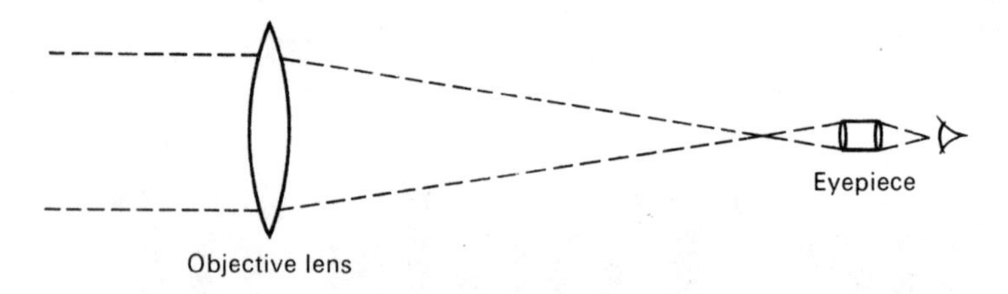

Fig. 3. A simple refracting type of telescope with a single element objective lens.

With this telescope Galileo discovered the four brightest satellites of Jupiter. The refracting telescope is so called because it refracts or bends the light to a point called the prime focus. However as the light passes through the lens it is split into its constituent colours or wavelengths, rather in the same way that a prism would do, and

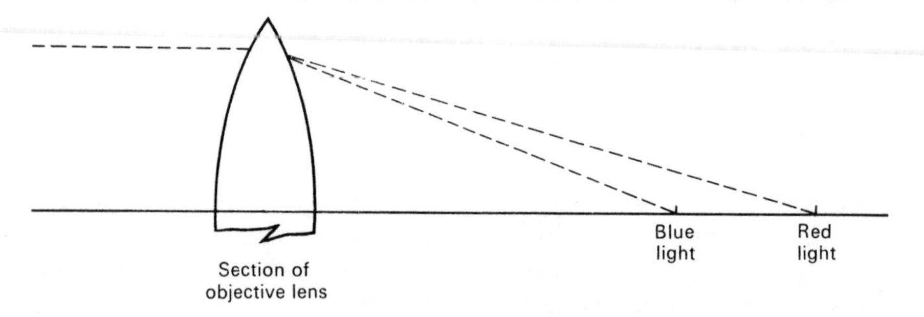

Fig. 4. Chromatic aberration bring different colours, or wavelengths, of light to different focal points by a single element objective lens.

causes the different wavelengths to focus at different distances from the objective lens, see Fig. 4. This can be overcome by either making the focal length very long, so reducing the angle through which the

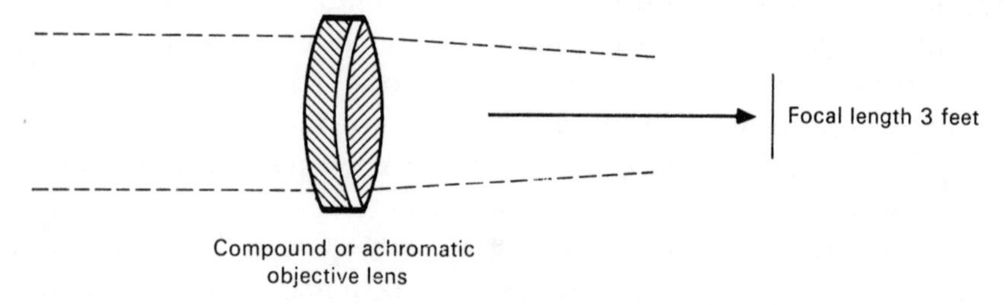

Fig. 5a. A two element or compound objective lens. Chromatic aberration is reduced by this method, using two different types of glass.

refraction takes place, or by making the lens from two different kinds of glass, each with a different refractive index as shown in Figs. 5a and b.

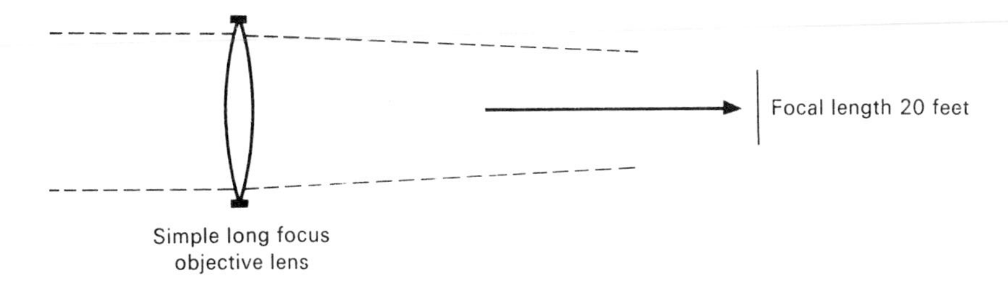

Fig. 5b. Chromatic aberration is much less apparent in a long focus lens because the light is refracted through a relatively small angle.

The effect that a lens has on light passing through it is known as Chromatic Aberration. Before the method of making compound objective lenses was perfected, astronomical refractors reached astronomical proportions. Some of the telescopes were built with focal lengths of 20 feet or more to obtain a better image.

It was Isaac Newton who realised the problem and started to find a way around it. Newton observed that a mirror could reflect light without any aberration at all. Using his newly developed 'fluxions' or calculus he designed a mirror which would act as a means of bringing the light to a focus in a short distance, without incurring any aberration. The mirror would have to be of parabolic section, reflecting the light back along the tube to where a small flat mirror placed at 45° to the axis of the telescope would bring the light to an eyepiece outside the tube.

After experimenting with many types of metal, Newton settled for an alloy of one part tin and four parts copper, called speculum metal. It could be cast easily and ground and polished to a perfect surface, and of course it was highly reflective.

Reflecting telescopes caught on, despite the fact that they had to be frequently re-figured to remove the oxidised layers. Years later came the technique of depositing a layer of pure silver chemically on a glass surface. This lent itself ideally to telescope mirrors as glass is easy to grind to a given surface and can be polished with an accuracy extending into millionths of an inch. The glass disc was polished to the required shape and a layer of silver became the reflective surface. The basic form of a modern Newtonian telescope is shown in Fig. 6.

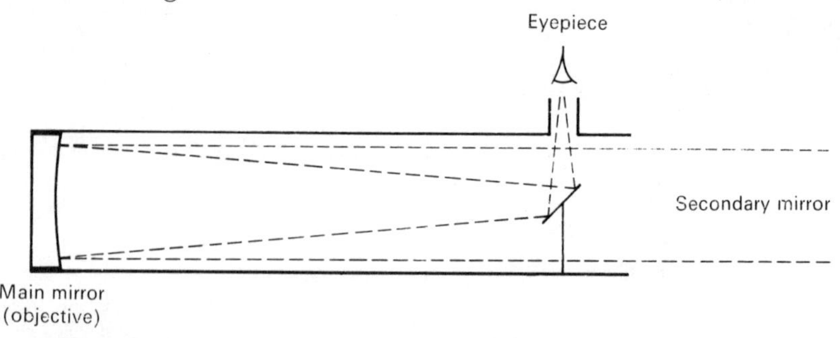

Fig. 6. The basic design of the Newtonian reflecting telescope has changed very little since its invention in the seventeenth century.

In later years it has become possible to use not silver as the reflecting surface, but aluminium. The aluminium is vaporised in a vacuum chamber in which the cleaned telescope mirror is placed. The vapour condenses on to the mirror surface. This so called aluminising process has several advantages. After the aluminised mirror is exposed to the atmosphere, a layer of aluminium oxide forms on the exposed surface. The oxide of aluminium is an extremely hard transparent substance, which after forming gives a protective coating on the outside while the deeper layer of aluminium is left in its pure unoxidised state.

In addition to the Newtonian type of reflecting telescope many others have been developed over the years. One of the best known examples being the Cassegrain system, shown in Fig. 7. There is a hole in the centre of the mirror through which the light can pass after being reflected from the elliptical secondary mirror. The

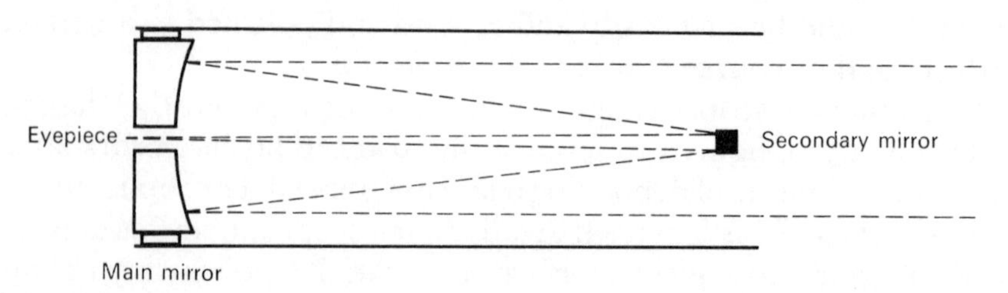

Fig. 7. The Cassegrain telescope is an all reflection system which can give a large focal ratio and very high magnification.

eyepiece is located behind the main mirror, in the same general position as in a refracting telescope. This type of telescope with its folded light path can give a highly magnified and sharp image, for this reason these telescopes are found mainly in the hands of planetary observers where this function is of primary importance.

Most of the different types of telescopes are now variations of this theme, incorporating one or more of the following basic components; objective lens, objective mirror, plane or flat secondary, elliptical secondary.

The first of these combinations is a very useful telescope as it uses an objective mirror as the main element. At the top end of the telescope tube is a Cassegrain type elliptical secondary mirror which returns the light towards another, plane mirror positioned at 45° to the optical axis bringing the light out to the side of the tube to

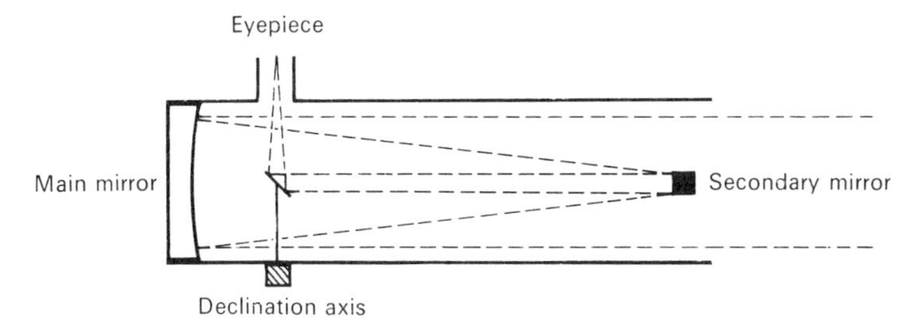

Fig. 8a. Combination telescope bringing the light out at the declination axis.

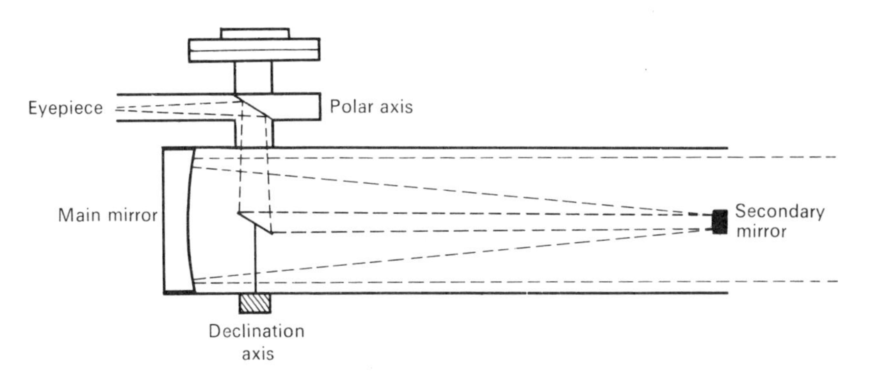

Fig. 8b. Telescope working on the Coude or Springfield system bringing the light to a focus along the polar axis. With this method the eyepiece remains stationary.

the eyepiece. There are two types of telescope within this group, the first uses an eyepiece situated in the declination axis, the second type has the light directed along the polar axis to the eyepiece, see Figs. 8a and 8b.

The last type of telescope which the amateur is likely to come across during his experience, is the catadioptric compound lens-mirror system. This is probably the ultimate in present-day optical systems, combining the lens from a refracting telescope and the mirror from the reflecting telescope. There are two names which are likely to appear, the Maksutov and the Schmidt. The first type, the Maksutov employs a concave-convex meniscus objective lens and a spherical objective mirror as opposed to a paraboloid. The meniscus lens acts as a correcting lens, to offset the effects of a spherical mirror, see Fig. 9a. The Schmidt telescope is basically the same as a Maksutov, the difference being that the Schmidt is solely used as a camera. Instead of bringing the light outside the tube to be observed with an eyepiece in the conventional way, a photographic film is

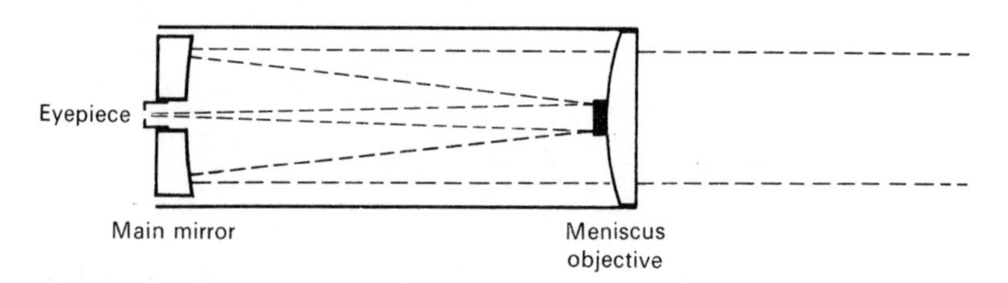

Fig. 9a. The Maksutov 'catadioptric' system combines both the refracting and reflecting principles into one telescope.

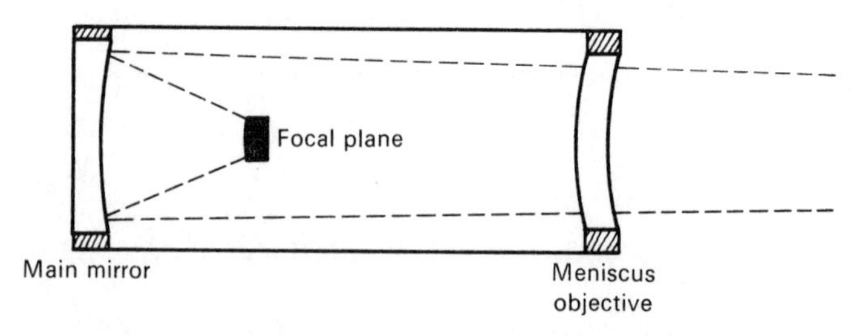

Fig. 9b. The Schmidt 'catadioptric' system is limited to use as an astronomical camera. The film is placed half-way along the telescope tube, an almost impossible position for physical observation.

inserted into the focal plane of the telescope inside the tube, see Fig. 9b. The Schmidt telescope cameras are used for deep sky photography where the maximum amount of detail is to be recorded, usually for the purpose of producing star atlases like the Palomar Sky Survey, which records stars down to the 21st magnitude and the Southern Sky Survey, carried out with a 48 in. (122 cm.) Schmidt camera at Siding Spring, Australia reaching the 23rd magnitude.

As already stated, the main use of the telescope is to gather as much light as possible and to bring it to a focus so that the eye can examine an image. Hence the primary consideration is one of light gathering power, in other words, diameter of objective lens or mirror. The next consideration is the magnification that a given optical system is capable of attaining. Magnification is a function of the focal length of the primary objective, the distance between the objective and its focal point, and the eyepiece which is used. Primary magnification or the magnification obtained through the objective alone is based on the shortest distance of distinct vision of the average human eye. Assuming the eye to give a magnification of 1 ×, the shortest distance of distinct vision is about 10 in. (25 cm.) therefore a telescope of 20 in. (50 cm.) focal length will give a primary magnification of 2 ×.

For a practical example, if we take a 10 in. (25 cm.) diameter telescope working at F10, the focal length will be 100 in. (250 cm.). The primary magnification will be 10 ×, but for the eye to 'see' the image an eyepiece must be used to present a real image. The magnification obtained by using an eyepiece can be expressed in terms of the focal length of the telescope, divided by the focal length of the eyepiece. In the above example, the focal length of the telescope is 100 in. (250 cm.) and if we use a 1 in. (2·5 cm.) focal length eyepiece, the total magnification will be 100 ×, for a ½ in. (1·25 cm.) focal length eyepiece 200 ×.

The F number is the ratio of the focal length of the objective divided by the diameter of the objective. Hence a 6 in. (15 cm.) telescope working at F12 will have a focal length of 72 in. (180 cm.). A 4 in. telescope with a focal length of 32 in. (81 cm.) will be F8. Most refracting telescopes have a focal ratio between F10 and F15, to keep the chromatic aberration to a minimum. This means that compared with a reflector which may be as low as F4, the refractor will have a high magnification. One of the considerations to be taken into account when deciding on the type of telescope to buy is the focal ratio. If one wishes to observe the planets, then a high F number is beneficial, but for faint deep sky objects such as nebulae and galaxies the lower the magnification the better and the use of much lower F numbers is required. The reasons for this are simple, the higher the magnification the lower will be the brightness of the image as seen with the eye. The image being spread over a large area. The planets can afford this loss of light over the surface because there is plenty of light to play with. The much fainter objects on the other hand cannot afford any light loss at all or they disappear completely.

With regard to the comparison of the types of telescope, a 6 in. (15 cm.) F10 refractor will under most conditions outperform a 6 in. (15 cm.) F10 reflector. A 6 in. (15 cm.) refractor will usually perform as well as a 10 in. (25 cm.) reflector under average conditions, and it is only when the conditions are perfect that the reflector will reach its theoretical limit. Pricewise, refractors cost very much more than reflectors. For the price of a 6 in. (15 cm.) refractor one can buy a 12 in. (30 cm.) or 14 in. (35 cm.) reflector. For this reason telescopes larger than 4 in. (10 cm.) aperture, in amateur hands, all tend to be of the reflecting type.

Another consideration, which can be directly linked to the magnification, is the field of view – the area of sky which is actually seen through the eyepiece. In general the higher the magnification, the smaller the field of view, while the converse is also true. For general observation, scanning the milky way and hunting for faint star clusters, a low power should be used. Once an object is located then a simple change of eyepiece will give a higher magnification, for studying the object in detail. For very low power I use a 60 mm. Kelner eyepiece to give magnification of about 27 × and a field of about 2°. There is one point to remember here though – with a low power the sky background will appear to be brighter and this effectively reduces the magnitude limit of the telescope and causes the fainter objects to be lost against the sky. For this reason a limit can be placed on the magnification that can be used effectively. The lower limit for most telescopes would be about 2 × per inch of aperture. The upper limit, depending to a small extent on the quality of the optical system can be from about 25 × per inch to 50 × per inch of aperture.

It is very difficult to state that a given telescope is required to observe any particular object. The best way is to look at a range of telescopes and compare these with some practical situations that can be found in the sky. A telescope is capable of separating objects that are close together and it is this that is the deciding factor as to whether the instrument will do the job required of it. The technical term for the ability of the telescope to separate close objects is its resolution, a numerical quantity pertaining to the closest objects that a given telescope can separate at its resolving power. To find the theoretical resolving power of any telescope including a radio telescope, the wavelength of light or radiation must be taken into account. The exact formula is;

$$R = \frac{1 \cdot 22\lambda}{D}$$

where λ is the wavelength of the radiation and D is the diameter of the objective in the same units as the wavelength. For optical telescopes this can be simplified as;

$$R = \frac{4 \cdot 5}{D}$$

where D is the aperture in inches. Table 1 gives telescope apertures and the resolving power. The magnitude listed in the table will be discussed in a later chapter but for now it will be sufficient to say

Diameter	in.	2	4	6	8	10	14
	cm.	5·1	10·2	15·2	20·3	25·4	35·6
Resolving power, Arc seconds		2·25″	1·12″	0·75″	0·56″	0·45″	0·32″
Magnitude limit		10·2	12·0	12·5	13·2	14·0	14·8
Linear miles		2·7	1·3	0·98	0·67	0·54	0·38
Resolution at distance of moon							
Linear km.		4·3	2·1	1·58	1·08	0·87	0·61

Table 1. Telescope capabilities with increasing aperture. Line 1, diameter of aperture. Line 2, resolution in seconds of arc. Line 3, magnitude limit. Line 4, diameters of the smallest craters that will be seen at the surface of the moon.

that it is obvious that the larger the objective diameter, the fainter the magnitude limit.

It must be stressed that the figures calculated in this way will give only the theoretical maximum capabilities. Do not expect your telescope to reach these limits. If it does then the optics are flawless and the weather conditions perfect. Most telescopes will reach to within 90% of these figures.

Most telescopes, if bought with eyepieces, will come as a package unit ready to use. The kit will contain possibly three eyepieces, two filters, one for the sun and the other for the moon. The better sets often contain a Barlow lens.

A Barlow lens is a device for effectively increasing the magnification of an optical system by increasing the effective focal length, see Fig. 10. The Barlow lens simply reduces the angle of the convergent rays of light entering the eyepiece.

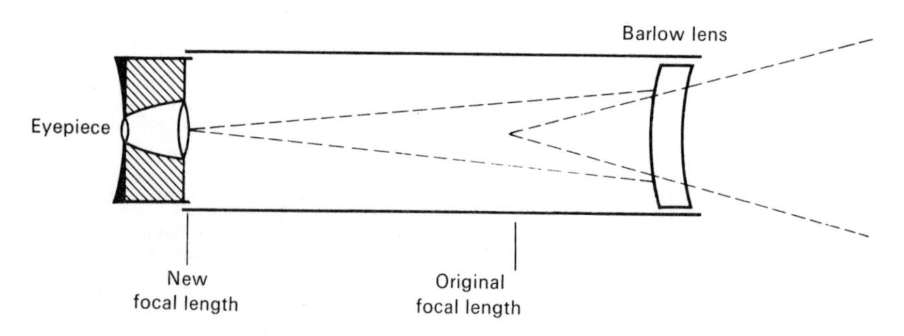

Fig. 10. The Barlow lens effectively increases the focal length of a given telescope by a set amount. The normal form of Barlow lens comes in 2× or 3× sizes, respectively doubling or trebling the magnification from a given system.

The moon filter, though not really needed with small telescopes, is a great benefit to larger instruments especially when using a low power, when the image presented to the eye can be very bright.

The sun filter is not a good piece of equipment though it will be safe enough if bought with the instrument. Never under any circumstances try to make any type of filter with the intention of observing the sun through it. The concentration of solar heat through even the smallest telescope is intense and the direct temperature at the focal plane may reach many hundreds of degrees Fahrenheit, which will shatter even good pyrex and burn through celluloid in much less than a second. If this image is allowed to fall on to the eye permanent blindness will result. The correct method of making solar observations will be discussed in detail in a later chapter. Before carrying out any solar observations please read through the relevant section in Chapter 5.

This equipment covers the basic needs of the beginner but much more expensive items can be purchased for specialised observations. One of the most common instruments used in conjunction with the basic telescope is, of course, the camera. The beauty of having the facilities for taking photographs of selected objects is that there is a permanent record of the observation; 99% of the time spent on observing in a professional observatory is taken up by exposing a photographic plate to the light of the object in question. One of the advantages of the photographic plate is its ability to accumulate the light which falls on to its surface over a period of time.

One of the most important parts of the telescope is the mounting. A bad mounting makes a bad telescope, no matter how good the optical components may be. There are two basic types of telescope mounting – the Altazimuth and the Equatorial. The Altazimuth mounting is the simplest type and is usually found on small commercial telescopes and camera stands. This did not mean that the mounting can only be used on small telescopes, though. The mounting has one axis perpendicular to the observer's horizon, which gives movement in azimuth, and the other axis at 90° to this to give movement in altitude, see Fig. 11. Although the altazimuth mounting is scorned by some as being inferior to the next type, it can be built to give a stable, vibration-free mounting. Arguments for this mounting are its simplicity to construct, its ability to take very large loads, its all sky accessibility and its cheapness. Arguments against include its unsuitability as a photographic mounting and that to keep a star in the eyepiece it is necessary to move both of the axes unless one lives at the south or north pole, when the telescope will work as an equatorial mounting.

The equatorial mounting is a more advanced type of mounting that needs a great deal of time and patience before it can be set up correctly. This type of mounting has one axis which must be aligned accurately parallel to the Earth's axis, called the polar axis. It can

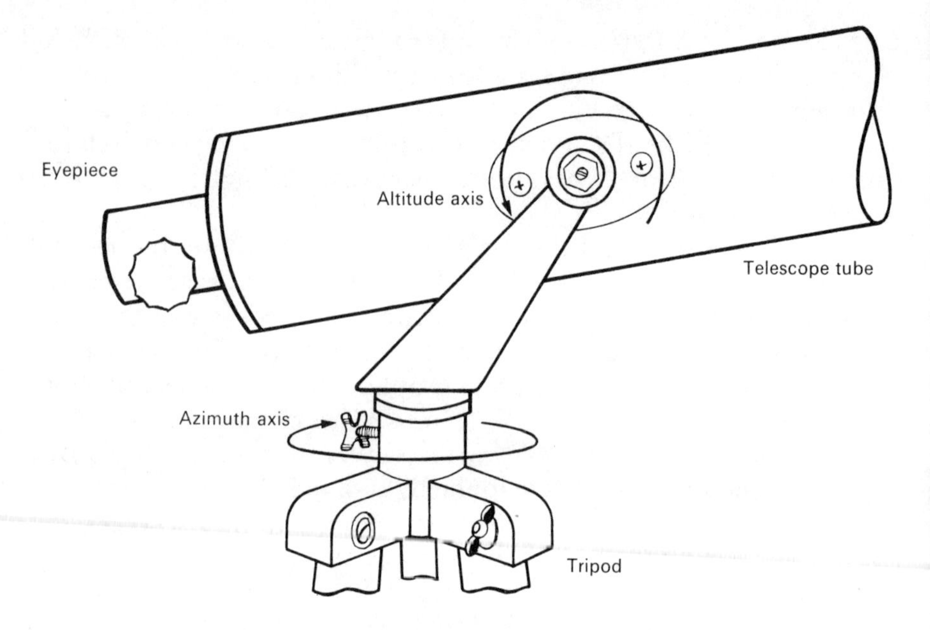

Fig. 11. The altazimuth telescope mounting is so called because of its axes of rotation, one in altitude and the other in azimuth.

be driven either manually or by a small electric motor to counteract the rotation of the Earth. The second axis, which is called the declination axis must be exactly at 90° to the polar axis. See Fig. 12.

Because of the ease with which the equatorial mounting can be made to follow the motion of the stars it has become the ultimate in telescope mountings. Advantages of this type of mounting are the ease with which the telescope can be made to follow celestial objects and its adaptability for photography and other uses where long periods of observing are needed on one particular object. Its disadvantages include the difficulty in setting up an instrument for accurate work, a point which is very important if the instrument is to be portable. For large telescopes the mounting needs to be very massive to maintain its stability and the cost can be quite high.

Once one has obtained a telescope of some kind and got used to its working limits, one can start to think about some of the auxiliary equipment that can be fitted to improve the value of the observations carried out. The first piece of equipment that we shall discuss

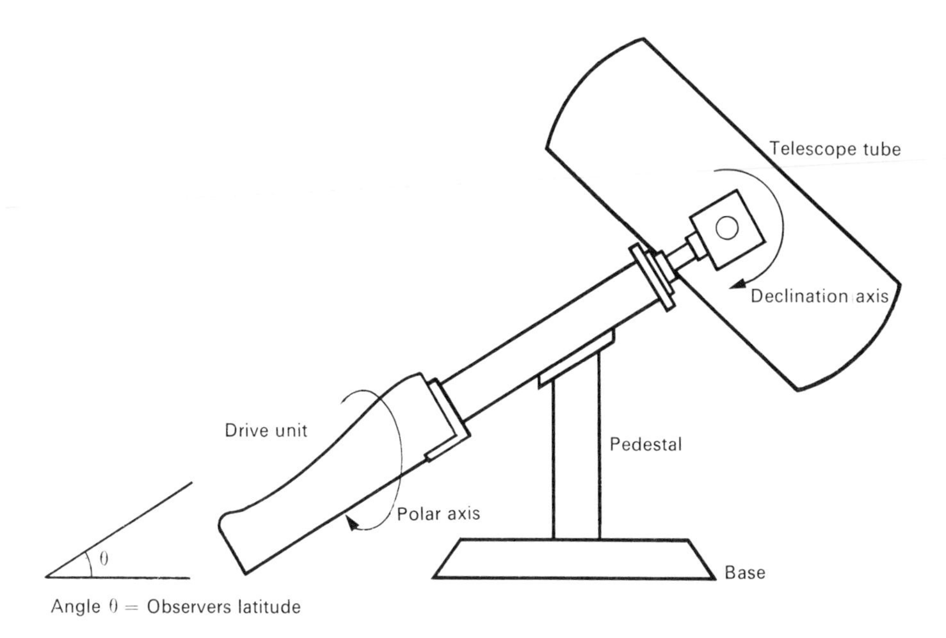

Fig. 12. Schematic diagram of an equatorial telescope mounting. The polar axis must be accurately aligned with the Earth's axis to give true equatorial movement.

is the micrometer eyepiece. This eyepiece has been developed to enable the accurate measurement to be made of double star separation and planetary diameters. The eyepiece has at its focal plane a system of cross wires, the traditional two fixed wires and a third wire which can be moved. The third wire is fastened to a micrometer enabling the distance between it and the fixed wire to be measured. Also for use with an eyepiece are some kinds of photometer. These usually consist of a system of mirrors in the tube of the telescope to deflect a point of light into the eyepiece. The light comes from a small bulb controlled by a rheostat and connected to a voltmeter. The intensity of the artificial light is then compared with that of a star and adjusted until both are the same, and the voltage is read from the meter. A fairly simple conversion scale is then needed to convert volts into magnitude. Although this equipment leaves much to be desired in the way of accuracy, it does the work effectively and with a good conversion table can be reasonably accurate. The problem is that the eye is not as sensitive to some colours as it is to others, and a scale is needed for a number of set stars of which the surface temperature and hence the colour is known. Hence the star to be measured is first given a colour class before the reading is taken. Red stars are measured on a different scale from yellow stars, yellow stars are measured on a different scale from the blue stars and so on.

To obtain a direct measure of the star's surface temperature a piece of equipment called a spectroscope is used. The spectroscope splits the light from a star into its constituent wavelengths and shows a spectrum of the light to the observer. The spectrum is seen to be crossed by several dark or bright lines, which are produced as the light passes through the atmosphere of the star. Each line is characteristic of a single element in that atmosphere. Hydrogen produces four main easily visible lines in the spectrum of our sun. Helium, calcium, magnesium are among the elements detectable in some stars. The basic components of the spectroscope are shown in Fig. 13. They are, a collimating lens and slit arrangement to pro-

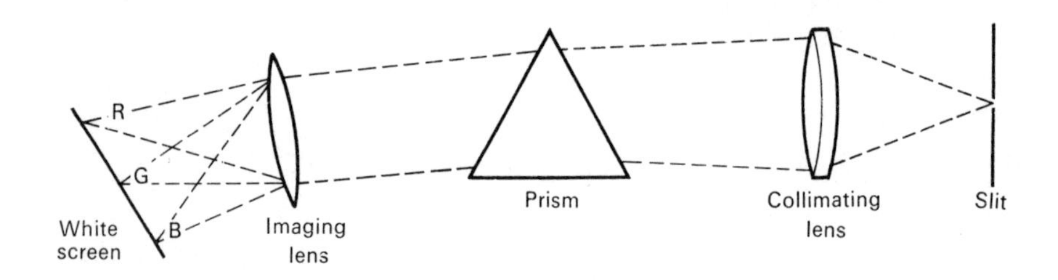

Fig. 13. A simple spectroscope that can be made from very cheap components. The slit should be about $\frac{3}{1000}$ in. (0.075 mm) wide and can be constructed from two razor blades.

duce a beam of parallel light, the prism to disperse the light into a spectrum, and an eyepiece or viewing screen to observe the spectrum. This type of simple spectroscope, although very crude, will show the basic principles and can be made from common components available at many opticians' shops, but a hunt around secondhand shops will be useful. Old prismatic binoculars, camera lenses, etc. are a source of cheap lenses and eyepieces.

The practical spectroscope differs slightly from the theoretical design in Fig. 13 in that to obtain a higher dispersion the prism may be made from a number of smaller prisms as in Fig. 14.

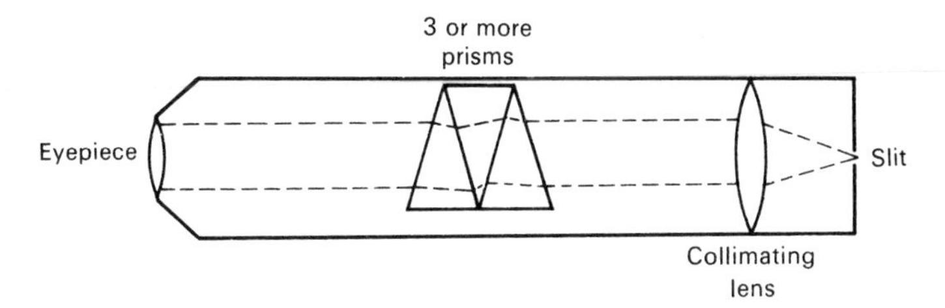

Fig. 14. A more complicated, direct vision spectroscope. The prisms have to be made from special types of glass.

A variation of the spectroscope is incorporated into the spectrohelioscope. In this instrument two slits are used, one before and one after the dispersing element. The element may be a piece of glass or aluminium with a series of lines ruled on the surface at the rate of about 15,000 per inch, called a grating. The slits cause the light to vibrate in the plane of dispersion. The instrument is focused on the Sun and the second slit placed over one of the dark absorption lines. On bringing a fairly high-powered telescope to bear on the second slit, an image of the Sun, or a portion of the Sun will be seen in the light of the absorption line in question, i.e. hydrogen or calcium. For example, the picture obtained in the light of hydrogen will consist of an image of the Sun in a light of wavelength range of only one or two Angstrom units, but where hydrogen is emitting light, the area will appear bright and where hydrogen is absorbing light the area will appear dark. Solar flares are clouds of very hot hydrogen usually above the normally visible surface of the Sun, emitting the characteristic light which is not normally seen, but the spectrohelioscope will show these flares as bright areas against the dark background.

Any branch of spectroscopy is complicated and requires some rather expensive equipment to carry out useful observations. For this reason I leave further explanation of the techniques and the necessary equipment to books which are solely concerned with the topic.

Astrophotography is a part of the study of astronomy that can be carried out with very modest equipment. I can only give here a brief outline of the techniques used to take photographs of some of the different objects that are encountered in astronomy, leaving the more detailed intricacies for the reader to experiment on and develop. Basically there are three areas of interest to the beginner; direct camera techniques, focal plane photography and eyepiece projection photography. Direct camera techniques involve guiding a camera to follow the stars to give a photograph of a selected area of the sky. As an experiment, place an ordinary camera focused at infinity, to point at an area of the sky where there are plenty of bright stars. Using a cable release to minimise vibration, take a series of exposures ranging from a few seconds up to several minutes. On the same film a second experiment can be carried out. Strap the camera on to a telescope that is equatorially mounted. Take as before several exposures but this time use the telescope to guide the camera as carefully as possible. For these experiments any camera will do as long as it has the facility for taking a time exposure and can be fitted with a cable release. For focal plane photography, it is beneficial to have the use of a single lens reflex camera, to make the work of focusing a lot easier. Remove the eyepiece from the telescope and point the telescope towards the moon. With the camera lens removed place the camera where the eyepiece would normally fit and through the camera focus the image of the moon as well as possible to give a sharp image. The exposure time in this instance will vary from telescope to telescope, depending on the primary magnification or focal length. Use exposures ranging between 1/500 and 1/60 second for the full moon. This is focal plane photography, using only the primary objective of the telescope as the camera lens, but to obtain a higher magnification it is possible to use an eyepiece to project an image of say the moon or a planet on to the photographic film. The techniques here can vary but it is usual to use an eyepiece in preference to the camera lens and it is this technique that I will explain. Again it is preferable to use a single lens reflex camera to aid in focusing an image. Place a fairly low power eyepiece into the telescope and focus the image. Now place the camera, without its lens, as close to the eyepiece as possible without touching it and re-focus through the camera. Exposure times again will vary with magnification but a range of exposures should be tried from say 1/125 to 1 second. With the last two techniques it is necessary, to obtain good photographs, to use a specially made mount which will hold the camera at the correct position to the eyepiece. These can be purchased with the telescope and if it is intended to be of primary importance to take photographs, then this simple and quite cheap piece of equipment should be included on the list of accessories to be purchased with the unit. With regard to the film that should be used, every observer has his own preferences to types and speeds. Basically, for the first experiments a fast black and white film should be used, but for lunar and planetary photography then a fairly slow film should be used, this will give a much finer grain and so will record more detail. One of the most important rules to stick to with astrophotography is never to be satisfied that one has found the perfect method, always be on the look-out for different ideas and techniques and experiment with as many as possible of these. Never be satisfied. Another way that one can experiment to obtain better results is to do the developing and printing of photographs at home. Although a moderate outlay is required to start with, the results obtained can soon show the advantages of having total control over the final product.

3 The Co-ordinate Systems and Time

Where is Paris? This question can be answered in a number of ways. Paris is 220 miles south-east of London, but this does not really tell us where Paris is, unless we know where London is situated. For example on a blank globe, with the information given above, it would be impossible to say or show where Paris, London, New York or Sydney are located.

This problem can be overcome quite simply, given one or two bits of extra information. To put the situation into the correct context we must mark the globe with a set of co-ordinates to which the above cities can be referred. A line drawn around the equator, marked off into 360 portions or degrees, can be our starting point. Lines can then be drawn joining the poles of the globe, each coinciding with one of the degree marks on the equator. In addition to this, lines parallel to the equator again at degree intervals are used to give positions north or south of the equator. Our problem can now be made much easier. Given the zero line of the co-ordinates it is possible to incorporate on to the surface of the globe the correct positions of the above mentioned cities. London has co-ordinates of 0° Longitude, 51°30′ Latitude. Longitude being the equatorial divisions and Latitude being the polar divisions (parallel to the equator). The co-ordinates of Paris are 2° 20′ E. (East of London), 48° 50′ Latitude. New York 74° W. Longitude, 40° 43′ Latitude. Sydney 151° E. Longitude, −33° 47′ Latitude. The negative value for the Latitude of Sydney denotes that Sydney is south of the equator.

It will be noticed that in addition to the degree markings there are minutes denoted by a single dash. Each degree is divided into sixty equal portions called minutes of arc in circular measure. For very accurate positioning, the minutes can be further subdivided into sixty equal parts called seconds of arc, there being 1,296,000 seconds of arc in a full circle.

I wish now to turn from the co-ordinates used on the earth and investigate the implications of such a system in a slightly broader sense. It is easy to design a co-ordinate system, such as the one described above, but there must be some way of determining the zero point of the system and, in addition to this, being able to determine one's position by an accurate method at any position on the surface. With such a co-ordinate system it was necessary to have a centre from which the system could be calibrated. But how was this to work? It was known that if the time could be kept with sufficient accuracy it would be possible to obtain the exact position of, say, a ship in mid-Atlantic, with respect to a fixed reference point. The reference point and zero of longitude became London or, more correctly, the Greenwich Observatory. The Observatory was commissioned by King Charles II in 1675. The aim was to obtain accurate positions of the stars and to produce tables of positions for navigation purposes. From the Greenwich Observatory it was necessary to adopt a standard time system to which observations could be applied to calculate the required position. This became the now familiar Greenwich Mean Time or G.M.T., also called Universal Time or U.T. From here on all time measurements will be referred to as U.T., as this is the standard adopted to enable observations from anywhere in the world to be compared. To find out the U.T. from the standard time at any longitude simply add the Standard Longitude Time Difference to your own Standard Mean Time. For example, an observer in South Africa will have a Standard Longitude Time Difference, S.L.D., of −2 hours, so to calculate the U.T. supposing the S.M.T. to be 2.30 a.m.;

U.T. = S.L.T.D. + S.M.T.
U.T. = 2 hours) + 2.30
U.T. = 0 hours 30 minutes

One point to remember is that some countries make an allowance for Special Summer Time and daylight saving arrangements which can vary from the S.M.T. by as much as 2 hours, and this must be accounted for when calculating G.M.T. Time keeping is not as simple as it may at first appear. Some simple experiments will clarify the more important aspects that arise when trying to keep accurate time tables. The first of these will show the general movement of the earth in its orbit. All that is required are some pieces of wood and a few nails. Firstly drive a sturdy post into the ground in a position with as good a South view as possible for Northern observers and as good a North view for Southern observers. Secondly take a piece of wood about 12 in (30 cm.) long, at each end of which a nail has been added to form a kind of sighting arrangement as shown in Fig. 15. Apart from the addition of another sighting device on a second post this is all that is needed.

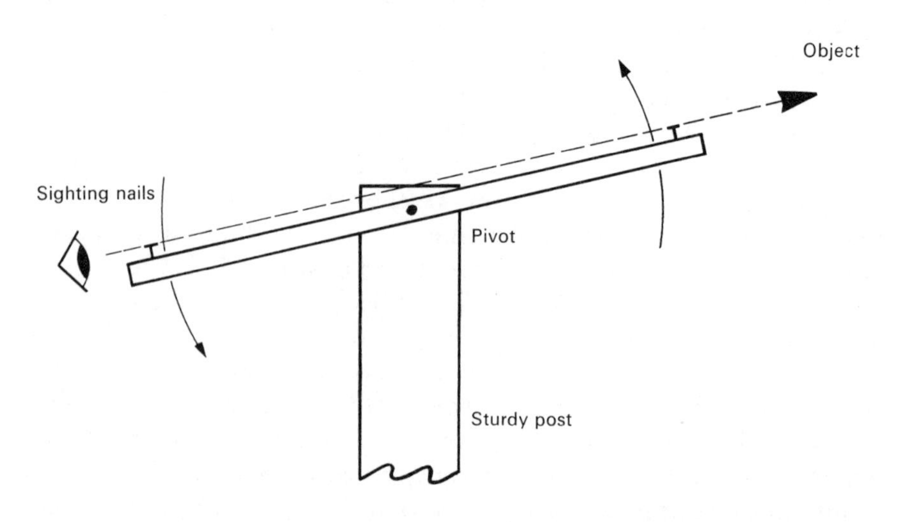

Fig. 15. A simple transit instrument. (When taking sightings of the sun do so by aligning the shadows of the nails on a screen – do not look directly at the sun!)

The observation is carried out as follows: about mid-day the instrument should be aligned so that the Sun is about to line up with the nails. As the Sun comes into alignment, make a careful note of the time. The more accurate the timing the better. Without moving the instrument this experiment should be carried out on successive days as often as possible and the results carefully taken down in the observation book. The most important point to remember is that while the observations are being made on no account should the instrument be moved or this will ruin the previous records. At the same time as these observations are being carried out a second instrument can be utilised to repeat the experiment with a bright star. These are known as transit timings. Accurate timings are usually taken as the object crosses a point which lies on a line passing through the Earth's poles and the point of observation, though for our purpose this is not necessary. I shall not relate the full results that are likely to be obtained with this equipment – this part is up to the reader to fill in. Basically the average time over exactly one year for the Sun to travel between successive transits is the Solar Day. It is this time by which we set our watches and clocks. The time taken between successive transits of the bright star is the Sidereal Day. The experiments above will show a slight difference between the two times. This is due to the fact that, as the Earth rotates on its axis, one revolution will bring the star back to its starting position, but the Earth has also moved round in its orbit slightly and to bring the Sun to its transit reference point it is necessary for the Earth to revolve slightly more than one complete revolution. This means that although the Earth has undergone one complete revolution with respect to the star, it has not quite completed one revolution with respect to the Sun, the Solar Day being a little longer than a Sidereal Day. Calculate the difference from your observations.

Problems do arise from this method of time and co-ordinate measurement system. It can only be used when the Sun is above the observer's horizon. The next step is to find a method which is suitable both during the daytime and at night. For this purpose the Greenwich Observatory was fitted with special instruments which were designed to work as transit instruments for the purpose of making observations of the declination of the bright stars, and relating the position of the stars to U.T. Basically the system is the same as that of the Earthly co-ordinates. An imaginary line extended through the sky above the Earth's equator forms the celestial equator and two points above the Earth's poles represent the celestial poles. The celestial equator, instead of being divided into degrees, minutes and seconds, is divided into hours, minutes and seconds. Each hour being one 24th part of a circle corresponding to the 24 hours in a day. This equatorial measurement is called Right Ascension (R.A.) and a star will move exactly 1 hour in R.A. during 1 hour of time. The celestial polar co-ordinates correspond exactly to the Earthly system in degrees, minutes and seconds, the celestial co-ordinates being called Declination, or abbreviated to Dec. Provided that the position of a star is known relative to the Sun, then the observer's position can be calculated with ease. In fact if a table of star positions is used this makes the work easier still.

From your own observations you will see that the stars are constantly changing position relative to the Sun. This is caused by the movement of the Earth in its orbit.

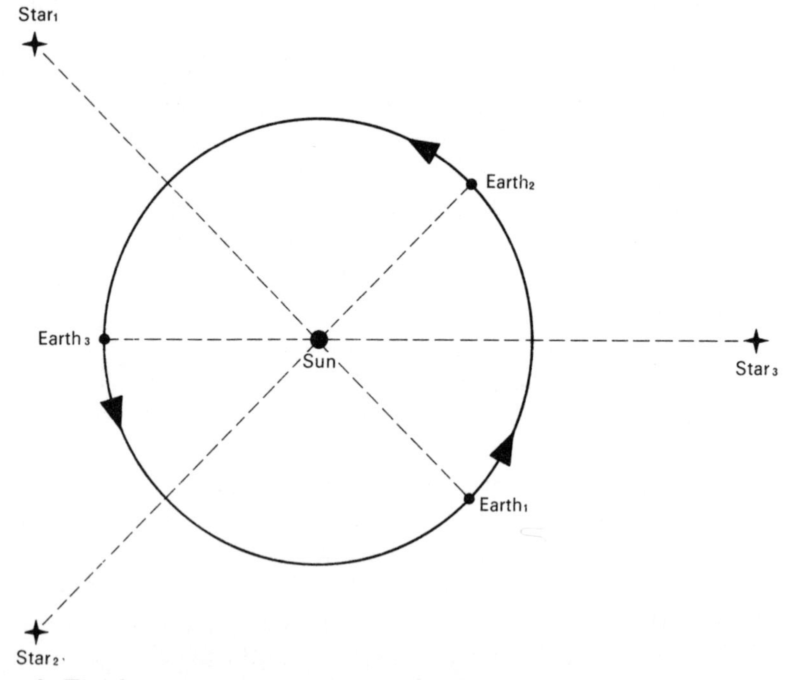

Fig. 16. The Sun appears to move against the star background as the Earth travels in its orbit. The path taken by the Sun is called the ecliptic.

This can be more easily understood by considering Fig. 16. As the Earth moves in its orbit around the Sun, the Sun appears to move against the star background – the stars that are visible at say midnight on a series of consecutive nights gradually change. It takes one complete orbit, or full year, for this complete cycle to take place. The path which the Sun describes through the stars is called the ecliptic, though this is really the path which the Earth takes in its orbit around the Sun. The ecliptic can be seen on the star atlas.

Notice also that the ecliptic does not follow the equator, it wanders both north and south of it. This wandering is caused by the tilt of the Earth's axis, so that the Earth's equator is not always in line with the Sun, see Fig. 17. Our seasonal changes are caused by this.

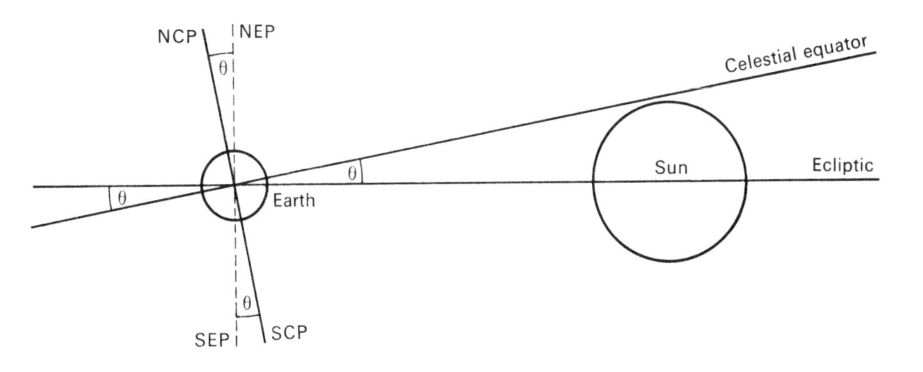

Fig. 17. The Earth's axial tilt in relation to its orbital path. The figure depicts the relationship during the southern hemisphere summer time (size and proximity exaggerated).

Twice per year the 23¼ degree tilt of the Earth's axis will bring the Sun above its farthest north and farthest south position on the surface of the globe. These northern and southern limits are termed the tropics of Cancer and Capricorn respectively, named after the constellations through which the Sun appears to pass at the two extremes of latitude. There will be two days in the year when the Sun will appear to cross the equator – once from south to north and again from north to south. These days are called respectively spring and autumn equinoxes, the terms spring and autumn being relative in the two hemispheres. The points at which this takes place are called the nodes. The Sun is at the ascending node as it crosses the equator from south to north and descending node as it crosses from north to south. From these points the co-ordinates of Right Ascension are calculated, the zero point of R.A. being at the point of the ascending node. Originally this point was called the first point of Aries because the node was situated in that constellation but because of a 25,000 year wobble in the Earth's orbit, known as precession, this point has now moved into the constellation of Pisces.

Precession, mentioned above, can be likened to the wobble given to a spinning top or gyroscope if touched while in rotation. The Earth's wobble being of a much longer period – 25,000 years in fact. Precession has the effect of changing the position of the co-ordinates of the celestial objects over a complete cycle of rotation, hence the change of pole stars over a period of time. Because of this effect, star atlases are only valid for a given length of time. All star atlases are marked with the period or year for which they are accurate. This is termed the 'epoch' of a star atlas. Normal amateur atlases can be used for up to 50 years before it becomes necessary to change to a more up-to-date atlas. This, of course, depends on the accuracy required.

For the beginner, the celestial co-ordinate system can be rather confusing with all of its intricate deviations from the norm. This brings us on to the last deviation that can be encountered, nutation. This is yet another small wobble imposed on the 25,000 year precessional wobble. The amount by which a star's position varies due to nutation is very small and only becomes important with very accurate measurements of an object's position. The beginner need not bother too much about this.

This then is the basic co-ordinate method used in astronomy. This can be employed to express the position of any spatial object with respect to the Earth's surface. It often becomes necessary to give a reference relative to another system and the first of these is the Solar System. The co-ordinate method used in the Solar System is usually called the ecliptic and enables the position of objects within the Solar System to be placed in a grid of reference points. The ecliptic, on which it is based, is the plane of the Earth's orbit. In three-dimensional geometry, this cuts the Solar System into north and south. The ecliptic plane is divided into degrees, minutes and seconds, and the north and south measurement is similarly divided. From the plane of the ecliptic to the south or north pole of the System is equal to 90°. The great advantage of this method is that it enables one to build up a three-dimensional picture of the Solar System, provided of course that the distance of an object from the Sun is known.

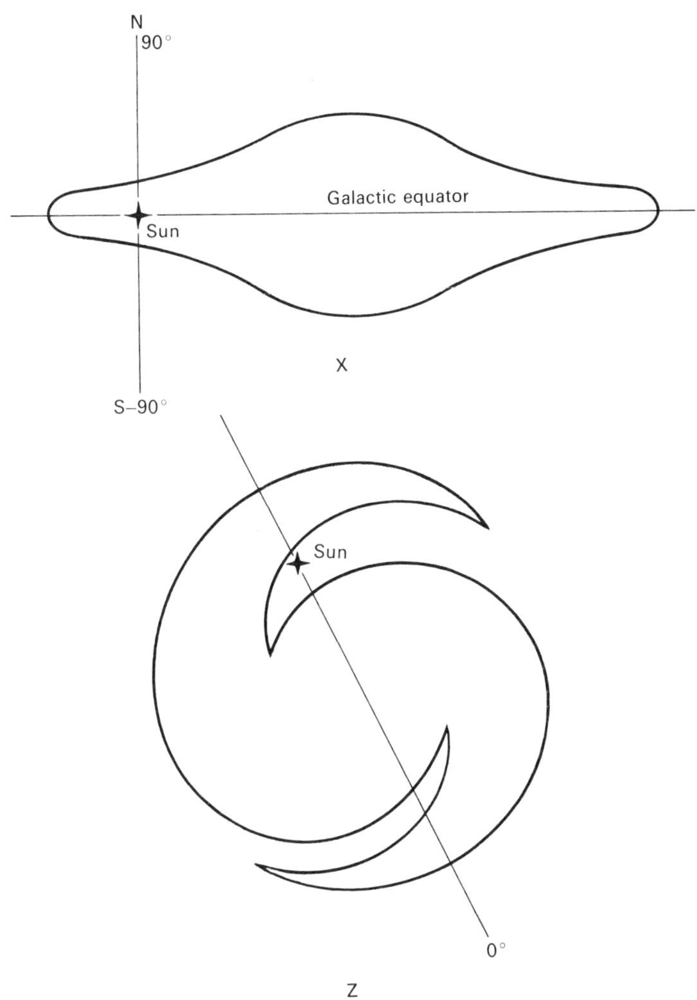

Fig. 18. The position of the Sun in relation to the galaxy and the galactic co-ordinates.

The other co-ordinate system that the amateur is likely to meet is that of the galactic co-ordinates. Basically similar in format to the ecliptic co-ordinates, the basis of the galactic system is the plane of the galaxy, or any galaxy for that matter, see Fig. 18. The galactic poles are the poles of the system. This system can be used to build up a picture of the galaxy in three dimensions. We will come on to the importance of this facility in a later chapter.

4

Making Observations

This chapter can only be used as a general guide to some of the methods in use. Each observer will find that with experience comes the development of technique. The aim of this chapter is therefore to point the observer in the right direction – from there on it is a matter of gaining the experience and developing that personal touch.

The most important part in making any observation is the method of recording the results. For this purpose it will be necessary to obtain a book which can be used to keep the observing record. The book should be fairly large and be pre-ruled, plain pages being added later if necessary. Plain pages are useful for the inclusion of drawings, to make the record more useful for future reference.

The first and most obvious entry in the observing record is the full date, which should include the time of the observation and the year. The time should be recorded as Universal Time or Greenwich Mean Time as this helps in comparing observations from anywhere in the world.

The second section should refer to the prevailing conditions at the time of the observation. This is important because while the naked eye, or a binocular or small telescope will not show much difference between a good clear night and a bad night, larger instruments are capable not only of magnifying the image of an object, but also of magnifying atmospheric turbulence by the same amount. The effect of this does not become apparent until one reaches say a 4–6 in. (10–15 cm.) diameter telescope. With smaller telescopes the effect of bad turbulences makes little difference to the image of the object under scrutiny. A large telescope will reveal turbulence as a slight wavering and shaking of the image and in severe cases the image may wander from side to side as much as 2–3 arc seconds, degrading the image to the extent that surface detail is completely obliterated.

The method used to record the type of conditions at the time of the observation have been refined over the years. The first scale used for this purpose was defined by the Italian observer Antoniadi and is shown in Table 2. Although there are five basic types of conditions encountered in the scale, some observers have at a later date tried to enhance this by dividing each of the Antoniadi definitions into two or more sections. This is really only making things more complicated and the simple scale is quite sufficient for most observing programmes.

I	Perfect seeing.
II	Some quivering of the image.
III	Much quivering, unstable image.
IV	Poor.
V	Very bad, hardly worth observing.

Table 2. The Antoniadi Seeing Scale.

In addition to the above estimates of the seeing conditions it may also be useful for reference purposes at a later date to include general notes about the weather. Turbulence can be caused by a large difference between night time and day time temperatures, high altitude cloud or ground mist. All of these seemingly irrelevant items of extra information can help to account for the nature of the seeing conditions and help the observer to become more aware of the causes and to be able to predict when good nights are likely to occur.

The next entry in the record is the observation itself. Basically what is required here is a record of the instrument being used, the magnification and type of eyepiece and any ancillary equipment. All of these help if it becomes necessary to repeat the observation and in analysing the information at a later date. The observation should be recorded in as full a detail as possible. At the time it may seem trivial to include some small detail, but it will pay off. Each observing record should be a full account of what was seen and should include any information at all that may help to carry out the experiment at a later date. All too often an entry will include ample information about what was seen, but when trying to repeat the observation with the same result, much time is wasted in finding the same technique that was used the first time.

Observing at night brings many problems of its own. On first going out of doors one will be in complete darkness. Gradually the eye will adapt to the amount of light that it will allow on to the retina and fainter objects will become visible. This is known as dark adaption and most eyes will adapt fairly quickly, within about 2 minutes of stepping out into the blackness. To gain full dark adaption may take quite some time, usually up to about half an hour but as soon as a bright light is switched on, say to read a star atlas, then the dark adaption is lost. For this reason all light should be kept to a minimum. Red light can be used in moderation as the human eye is not too sensitive to it. A suitable red light can be made by painting some dark red nail varnish on to the lens or glass of an old flashlight. Cold weather also has to be overcome when observing at night. One cannot observe comfortably with cold hands and feet, and a concrete floor will rapidly conduct heat away. One of the first steps to take against the cold, is to cover the observation site with an old carpet, some pieces of wood, or anything else that will afford insulation against the cold floor. Secondly, because heat rises, the body loses a great deal of heat through the head and so a thick woollen hat should be worn. Common sense is a good guide to any extra clothing that is needed, depending on the ambient temperatures at the time, and it is important to be as comfortable as possible at the observing site.

Some of the following items of equipment may be needed during an observing night – star atlas, notepad, pens and pencils. Any notepad will do for rough notes at the telescope. The best type of pens and pencils, I find, are those with a ring at the end which can be fitted with a loop of string and hung around the neck. For naked eye or binocular observation of overhead skies a deckchair or similar

eyepieces	pencils
red torch	clock
tools	camera
notepad	film

Table 3. Observing kit list. This should be checked before an observing session to make sure that all equipment that might be needed is present.

kind of reclining seat can be a great help. All equipment that might be needed for a night observing session should be kept in a box made for the purpose, suitably padded out to avoid damage to any of the pieces. Table 3 shows the list which is taped into the box in which my observing kit is kept.

One can now go out and carry out some simple observations. The first thing that one will notice is that the stars are not all of the same brightness. It is of course necessary to have some scale of brightness to which we can refer objects. This scale is known as the magnitude scale and works on the logarithmic principle, enabling a large difference in true brightness to be expressed as simple number. The magnitude scale works to the base 2·512 so that a difference of 5 magnitudes on the scale is equal to a difference in real brightness of 100. The brightest stars in the sky are of about magnitude 0, and depending on the conditions at the time, the faintest stars visible will be between magnitude 5 and 6. In the appendices is a list of the brightest stars. At the top of the list is Sirius, in the constellation of Canis Major. This star has a visual brightness of −1·47, the minus sign indicating that the star is brighter than magnitude 0. Stars fainter than magnitude 0 have a positive value. To calculate the real difference in brightness from two given values of magnitude, the following formula is used. If 'n' is the difference in magnitude,

$2{\cdot}512^{n}$ = difference in brightness

e.g. if the magnitude difference is 3·5 then

$2{\cdot}512^{3.5} = 25{\cdot}12$

This calculation is made easily with an electronic calculator, but it can be carried out using ordinary logarithms.

This magnitude scale can be used to express the brightness of any object in the sky, whether it be a star or a planet or even the sun. It is purely a measure of the amount of light that we, on Earth receive from the object. The brightness of the Sun expressed on the magnitude scale is −26·7, which means that it is 47,000 times brighter than a star of magnitude 0. −26·7 is a much easier way of expressing the difference. This scale that we have discussed is the apparent magnitude – it should not be confused with absolute magnitude. Absolute magnitude, often written as a capital M, takes into account the fact that the stars are at really immense distances. For example two stars, which may appear to be the same as seen from the Earth, may in fact lie at greatly differing distances. Absolute magnitude corrects the brightness of a star and expresses the brightness of stars as they would appear if they were all at the same distance. The standard distance adopted for this is 32·5 light years, the light year being the distance travelled by light in one year, 5·88 million million miles, (9·46 million million km.). Once we know the distance of a star it then becomes possible to determine its absolute magnitude and how it would compare with other stars at the same distance. The method of calculating absolute magnitude is quite easy with an electronic calculator.

$$\text{Absolute } M = \text{apparent } m + 5 + 5 \log \pi$$

$$\pi = \text{parallax in arc seconds} = \frac{3{\cdot}25}{\text{distance in light years}}$$

Parallax and stellar distances will be discussed in further detail in Chapter 7. The brightness of the Sun on this scale would be 4·8, a rather insignificant value when we consider that the brightest star in the sky has an absolute magnitude of −7·0, some 52,000 times more luminous than the Sun.

I have digressed slightly from the true objective of this chapter. Another problem that will occur when observing at night is the dewing of equipment. It cannot be overcome, so we must do everything in our power to alleviate the problem as much as possible. Dew forms as the Earth's atmosphere cools and reaches the saturation point, when moisture is deposited as small droplets on anything that has cooled down with the atmosphere. Under certain circumstances a small heater can be used to prevent dew from forming, but with a telescope used for visual work the heater causes too much air turbulence. Generally the visual astronomer is at the mercy of dew. There are some rules which must be mentioned here with respect to the optical system. Never wipe lenses that have dewed with any old piece of cloth. Cloth is one of the best collectors of dust and grit, and a lens or mirror can easily be scratched and damaged by this especially if it is damp. In general only clean lenses when it becomes absolutely necessary and then only use a cloth which is really clean, washed several times in warm soapy water and rinsed in clean, preferably distilled water. Never use a spectacle cleaner, as these often contain an abrasive which can take off the delicate coating which is applied to lenses to increase their transmission value.

The observing site can make more difference to the quality of the image than many people realise. Just 200 ft. (40 m.) or so in height, gained by travelling a short distance can mean the difference of a thick blanket of fog or mist and a perfectly clear sky. For this reason I promote the idea of travelling to a well chosen observing site, if not for all observations then most certainly for special observational topics where a clear sky is almost important, and for special occasions when a particular study is to be made. Many readers may live in or near to a town or city where a pure waste of energy can be observed every few yards in a large sodium light. This type of lighting can kill the image of a planet by brightening the sky background. Although sodium light is easy to filter out with special equipment, it is very expensive. So anyone with sodium light problems should look to the surrounding countryside for a better observing site, preferably as high above sea level as possible.

Those who are free from sodium lights should not assume that they have the perfect observing site – it may be better than some, but they should always be on the lookout for a better place. One important point here to remember is that permission must be obtained from the owner of the land before you use it, and if you do intend to use an external site it is often advisable to inform the local police force of your intentions with some idea of the time that you expect to be operating.

The first task that the beginner will want to set himself will be to learn the constellations. This primary learning period should not be passed too rapidly, the better one knows the constellations the more interesting will be subsequent observations and the easier it will be to find the fainter objects. Finding objects which are not visible to the naked eye relies to a great extent on a thorough knowledge of the stars in the area that act as guides to the object.

Once a good knowledge of the constellations is obtained, then one can proceed with a binocular or a small telescope. There are countless objects that can be observed with relatively cheap and simple optical systems. Variable stars are a good topic for a binocular observer, because a fairly wide field of view is required to enable as many of the comparison stars as possible to be observed without losing sight of the variable star itself. The idea is to compare the brightness of the variable star with some other stars which are of known brightness and to obtain an estimate of its magnitude. A full discussion of the types of variable star and the characteristics of the different types is included in Chapter 7. For the observer who wants to carry out a topic such as this should contact his national astronomical association who have special star atlases of variable stars with the magnitudes of all local stars that can be used for comparison.

Double star observing can be very useful, and this is one of the areas in which amateur work can be valuable to the professional observatories. This branch of astronomy does, however, require some rather expensive equipment and a moderately large telescope before really worthwhile work can be carried out.

Novae are dealt with in Chapter 7 where the full explanation of the objects is included. From the observational point of view, nova searching is a very useful pastime. Basically novae are exploding stars, they suddenly appear in the sky with no form of warning, becoming bright very rapidly. It is during this rise in brightness that the professional astronomers need to be able to turn the very large telescopes on to the phenomenon and examine it with the spectroscope. Very little is known about these stars and the more novae that can be observed the better, for obvious reasons. Nova spotting simply means that one hunts for these extra stars as they appear. A thorough knowledge of all the stars in a particular area of sky is needed, the area being periodically checked for nova stars. This can be carried out with quite simple equipment such as a binocular, which is the best instrument for the purpose. Nova searches should be confined to the plane of the milky way, where most of the novae are likely to occur, because of the greater population of stars.

Lunar and planetary observation can be very rewarding, but in these days of space travel much of the work has been taken out of the hands of the amateur. Drawing the Moon can be a satisfying project, possibly with the final aim of making a lunar atlas. Observing the planets requires rather larger telescopes and can become very specialised. The inner planets, Mercury and Venus, can be very elusive and because of their close proximity to the Sun are difficult to observe. Noting the phases of these planets and possibly calculating the orbital period from the observations can provide hours of enjoyment.

The outer planets – Mars, Jupiter and Saturn – can be very interesting to the amateur astronomer. Mars has its dust storms, Jupiter the Great Red Spot and the cloud belts, and Saturn of course has the beautiful ring system. These planets all require a fairly large telescope to show any detail, Mars especially so because of its small size. Jupiter's equatorial belts can be seen with a 3 in. (75 mm.) refractor or a 4–5 in. (100–125 mm.) reflector. Saturn's rings can also be seen with this size of telescope but to observe the Red Spot on Jupiter then at least a 3 in. (75 mm.) refractor or 6 in. (150 mm.) reflector is required. The other planets – Uranus, Neptune and Pluto – are difficult objects at the best of times. Only the world's largest telescopes will show any detail at all and even the 200 in. (5080 mm.) telescope on Mount Palomar will not reveal the planetary disc of Pluto. Uranus and Neptune can be found with a binocular if one knows where to look.

Fig. 19. The Pleiades star cluster, photographed from the Tattershall Bridge Observatory. The photograph reaches the 11th magnitude.

Star clusters of the open or galactic type can provide hours of amusement. Over a hundred of these objects are known, ranging from quite bright naked eye objects to the faint ones containing only a few stars. Some of the brighter examples are included on the star atlas, but a search of the milky way will often reveal others as misty patches of light almost at the limit of visibility. Among examples are, in the northern hemisphere, M45 and the Pleiades (Fig. 19) and, in the southern hemisphere, Herschell's Jewel Box in the constellation of Crux, NGC 4755.

The star maps can be found at the end of this book. Each map has the date and time of the month that the stars will take approximately the positions shown. It will be noticed that it is not practical to observe certain constellations at or during certain months, simply because that constellation is only visible during daylight hours. The observer will have to wait until the Earth is a little further around its orbital path to bring the constellation in question into position for viewing during the dark hours.

5
The Solar System

The Solar System is generally thought of as local space, the boundary of which can be defined in several ways. The most usual definition is that of the gravisphere. The major body in the Solar System is the Sun, which is an immensely large object and consequently has a great deal of gravitational influence, gravity being a function of the mass of the object. It is this gravitational field that holds the planets in orbit around it. The maximum distance at which an object can remain in orbit around the sun is about 100 thousand million miles (160 thousand million km.). Any object at a greater distance than this will not receive enough force to remain in orbit and hence will just float off into space, albeit on a slightly hyperbolic course. This then is the first and one of the more important boundaries – the maximum distance at which an object can remain in a stable orbit around the Sun.

The second limit or boundary is the distance from the Sun that radiation from the Sun is dominant. This is about 2½ thousand million miles (4 thousand million km.). At this distance it is really quite cold, with temperatures only just above absolute zero, about 4 °K or −259 °C. Although in theory radiation from the Sun, like gravitational forces, extend to infinity, in practice this is not quite so, because absorption of radiation by interstellar molecules soon reduces the radiation to zero value. Light energy is not affected to quite the same extent as heat energy or infra-red rays.

The visible limit of the Solar System is of course the planet Pluto, or until the year 2016 the planet Neptune, because for the time being until 2016 Pluto is inside the orbit of Neptune and closer to the Sun. It is quite possible that there is another planet, or more than one, outside the orbit of Pluto. Through distance and faintness it is unlikely that with present instruments we shall be able to find such a planet but there is always a chance.

Before going on to discuss some of the more pertinent aspects of the Solar System, it will be useful here to digress to some of the terminology used to express certain situations and configurations that occur.

Aphelion. The farthest distance from the Sun that a planet can be, in its normal orbit. This term can also be applied to satellites of planets where the planet's name is incorporated instead of 'helion', for example 'apogee' 'apojove' representing the Earth and Jupiter respectively.

Conjunction. The point in an orbit when a planet goes behind the Sun, as seen from the Earth. When dealing with an inferior planet, as the planet passes behind the Sun it is termed superior conjunction, or in front of the Sun inferior conjunction, see Fig. 20.

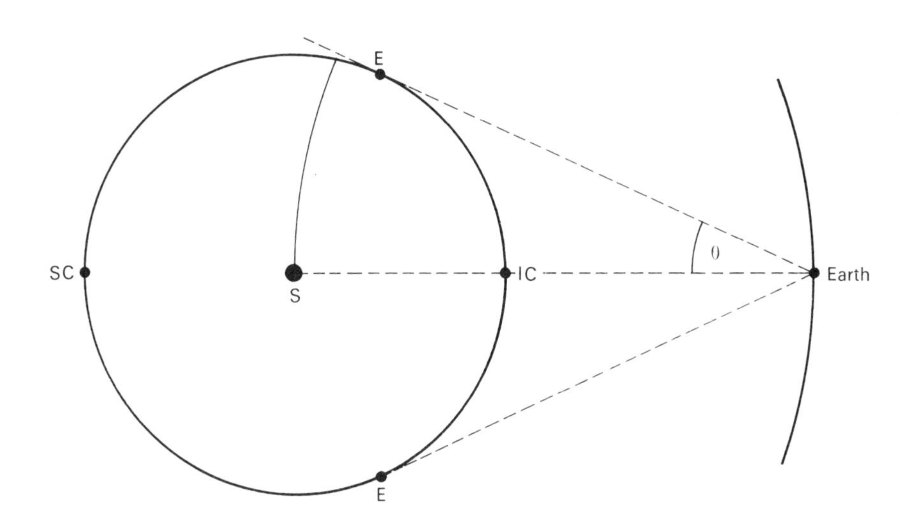

Fig. 20. Positions in the orbit of an inferior planet. S = Sun, E = greatest elongation, IC = inferior conjunction, SC = superior conjunction. Angle θ represents the planet's greatest observed distance from the Sun.

Diameter, Angular. The angle which the diameter of a planet's disc subtends from the distance of the Earth.

Elongation. The angular distance between the Sun and a planet as seen from the Earth.

Inferior Planet. Planet whose mean distance from the Sun is less than that of the Earth, i.e. Mercury and Venus.

Opposition. The point in the orbit of a superior planet when the Sun, the Earth and the planet are in line.

Perihelion. The point in an orbit that brings the planet in question nearest to the Sun, the opposite of aphelion. Again 'perigee' and 'perijove' refer to moons or satellites in orbit around the Earth and Jupiter respectively.

Sidereal Period. Period of an orbit or planetary rotation measured in relation to the fixed stars.

Superior Planet. Planet whose mean distance from the Sun is greater than the mean distance of the Earth, i.e. Mars, Jupiter, etc.

Synodic Period. Period between successive conjunctions, or oppositions of a planet.

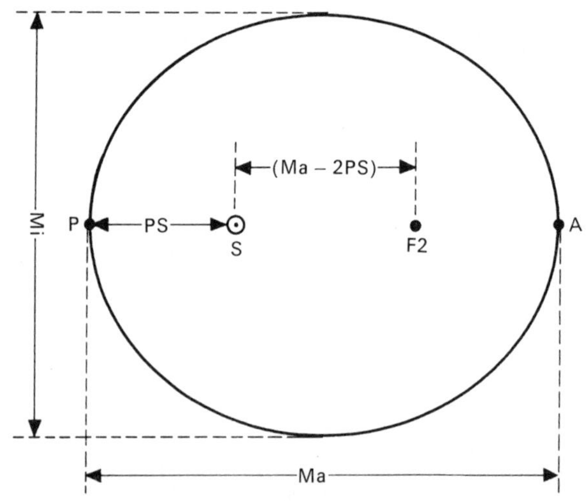

Fig. 21. Dimensions of an orbit. S = the Sun, F2 = the empty focus, A = aphelion point, P = perihelion point, PS = perihelion distance, Ma = major axis, Mi = minor axis, Ma − 2PS = distance between Sun and F2. The eccentricity of this ellipse is (Ma − 2PS)/Ma = 0.37878, approximately to scale.

All orbits are elliptical, the amount of ellipticity or more often the eccentricity being expressed as a number, see Fig. 21.

This means that, as a planet travels around the Sun in its orbit, at some point on the orbit the planet will be farther away from the Sun than at other points on the orbit. As the distance varies so does the gravitational influence of the Sun on the object, causing a change in the orbital velocity. The ratio of change in the orbital velocity can be calculated approximately from the relation,

$$\frac{P^2}{D^3}$$

where P is the period and D is the distance of the Planet from the Sun.

This formula breaks down when trying to compare two different orbits, partly because of the mass differencc between two different planets.

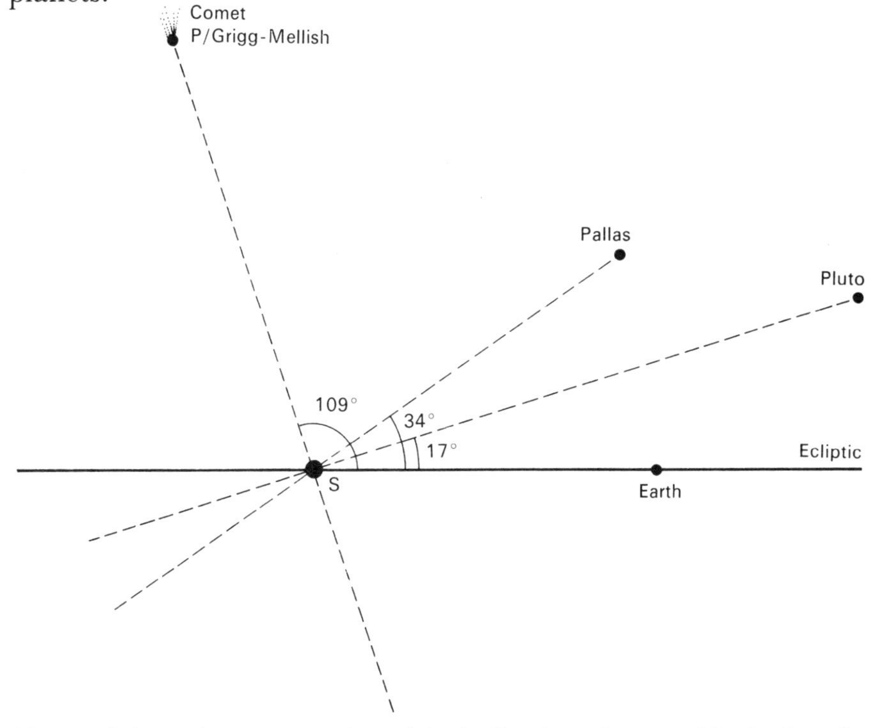

Fig. 22. Schematic representation of the inclination of some orbits in the solar system. Pallas is an asteroid and P/Grigg-Mellish is a periodic comet.

There is also a second parameter which must be taken into account when dealing with objects in the Solar System, and that is the inclination of the orbit. All of the major planets in the Solar System orbit in much the same plane as the Earth – only Pluto varies from this by any great amount. Therefore the major planets can all be found within a degree or so of the ecliptic, see Fig. 22. Some of the smaller bodies can have orbits which are inclined at very large angles to the ecliptic.

Of course there is much more to celestial mechanics than this, but these are some of the more important points. We now move on to the objects which go to make up the Solar System, beginning with the Sun.

The Sun

The Sun is the major body in the system, it provides the gravity to hold the other objects in place and provides heat energy to warm the immediate surrounding area. The Sun is really an immense body, its equatorial diameter is 0·86 million miles (1·4 million km.) and its mass about 2 thousand million million million million tonnes or more simply $1{\cdot}99 \times 10^{33}$ grammes. At 92 million miles (149 million km.) from the Earth it is by far the closest star to us, and from this its use as a test object for the theory of stellar evolution, which we shall come to in a later chapter, is of great importance. Our Sun like all other stars, shines brightly because it is very hot. The surface temperature of the Sun is about 5700 °K. This heat is radiated out from the centre, and therefore it is sensible to assume that the internal temperature is much higher than the surface temperature. Indeed it is, the core temperature of the Sun is about 12 million °K, but where does the energy come from? Well, at this temperature the main constituent of the Sun, hydrogen atoms, can be forced together to form a different type of atom, helium. During this, four hydrogen atoms combine to form one helium atom and some of the mass of the hydrogen is lost. The great Albert Einstein was the first to suggest that matter and energy were interchangeable by the formula;

$$E = M c^2$$

where E is the energy produced by mass M and c is the speed of light. By rearranging this formula, if we know the energy produced, simple maths will show that about 4 million tonnes of matter is being converted into pure energy every second. Please do not worry though, the Sun has enough mass left to carry on burning in this way for about 14 million million years or so at this rate. However changes will occur in the central regions of the Sun much earlier than this, making the above figure look rather silly. After a period of hydrogen burning, when the percentage of hydrogen gets too low to be suitable for nuclear reaction, then the central core of the Sun will increase in temperature until helium can burn, forming the elements carbon and nitrogen. The outer shell of the Sun during this phase will expand to giant proportions with the outer surface of the Sun possibly reaching as far as the orbit of the Earth. This is called the red giant phase, because the energy produced is radiated over such a large area that the surface temperature may drop to as low as 3000 °K. We shall later see that stars can be observed from the Earth that have these physical properties.

Observations of the Sun are not easy to carry out, mainly because of the amount of light and heat that comes from it. Even though this energy is dispersed considerably by the time it reaches the Earth's

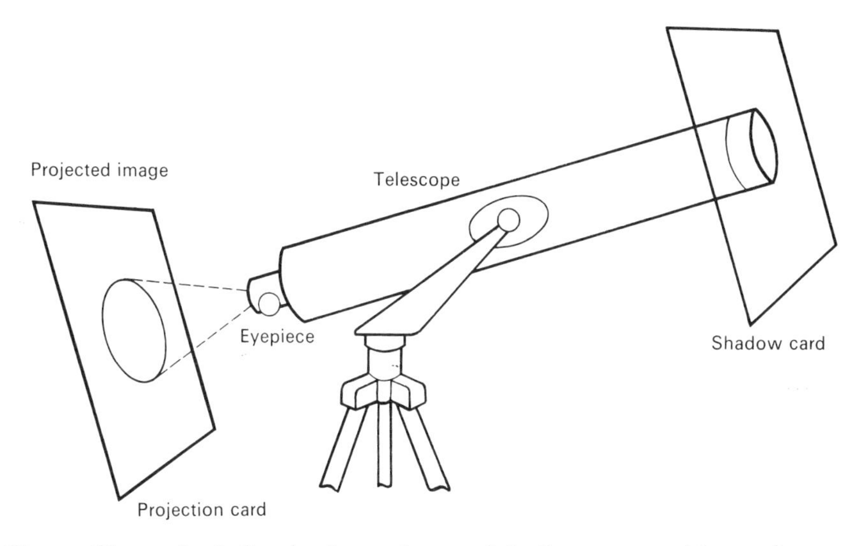

Fig. 23. The method of projecting an image of the Sun on to a white card screen. The shadow card is fastened to the telescope tube in order to cast a shadow on the projection card to give a better image of the Sun. The telescope is focused in the normal way.

surface there is still enough heat to do quite a bit of damage. The sunlight focused by a telescope mirror or objective lens has exactly the same effect as a 'burning glass' except that instead of burning holes in paper, the telescope could be focused into the eye with disastrous results. Instantaneous, permanent blindness would result the instant the narrow beam of light entered the eye.

The correct way to observe the Sun is by the projection method. This means that instead of projecting the image on to the retina of the eye, it is projected on to a sheet of white card, see Fig. 23. Try this experiment with a binocular. The card should be about 12–20 in. (30–50 cm.) from the eyepiece and focusing is carried out in the normal way. An image about 2 in. (5 cm.) in diameter can be obtained from a moderate 10 × 30 binocular in this way.

This method of observing the Sun is quite safe and different eyepieces used with say a 2 in. (5 cm.) diameter telescope will reveal quite a lot of detail. Even a binocular will reveal the dark spots, called sunspots, on the surface.

Sunspots are magnetic disturbances which show at the surface by causing turbulence. The reason that the spots appear darker than

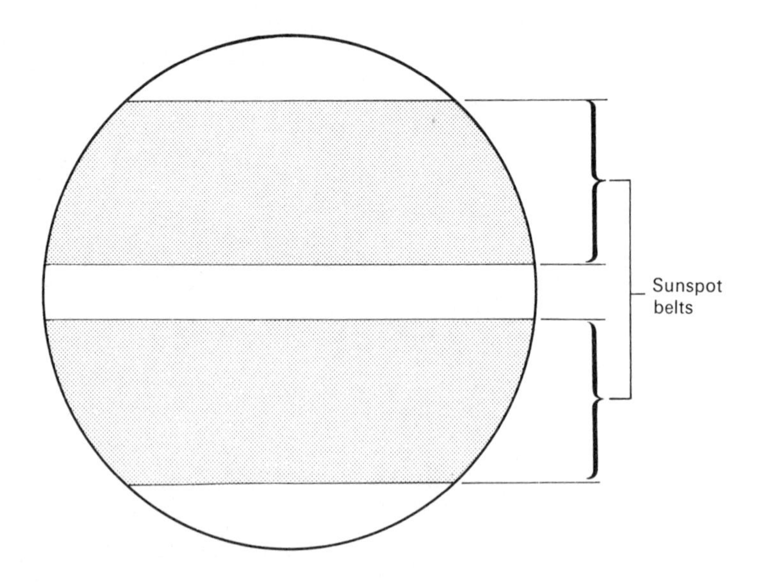

Fig. 24. The two sunspot belts. Although sunspots may occur outside this belt, these occasions are very rare.

the surrounding area is because they are much cooler than the surrounding gases by about 1000 °C. The exact reason that our Sun should be plagued with spots is not really known. Observation shows that the spots are linked closely to a cycle of maximum and minimum activity with a period of about 11 years, although there is some irregularity in the length of the period. The cycle begins with minimum, when there will be relatively few spots, but as the cycle progresses spots become more regular but are confined to narrow bands of latitude about 15° wide centred on 25° north and south latitude. As the cycle reaches its maximum, about 4 years after minimum, the spots can be observed in a wide band from 5° either side of the equator to 35° north and south latitude, see Fig. 24.

Closely associated with sunspots are some bright areas on or above the surface, called faculae. They are in some way connected with sunspots because faculae usually appear over the area where a spot is about to appear, this phenomenon often occurs some hours or even days before the spot is actually visible, the faculae then remaining for some days after the spot has completely disappeared. One observation that can be carried out using moderate equipment is the timing of the Sun's rotation period. This is done by observing the travel of a group of sunspots as they traverse the Sun's surface and plotting the daily movement. With luck one may catch a very large spot group that lasts one rotation period, making the calculation of the rotation period quite easy. Otherwise one has to catch a spot group as it emerges from one limb of the Sun and time it as it crosses the disc. Most of the smaller groups only last for a short time, usually less than one full rotation period, which makes evaluation of the period more difficult. One of the interesting points that has come from the observation of sunspots is that at different latitudes the Sun rotates at different rates. At the equator the period of rotation is about 25 days whilst at the north and south poles the period is thought to be nearer to 30 days.

Other solar phenomena are only visible at the time of total solar eclipse or with a spectrohelioscope discussed earlier. At the time of a solar eclipse, the corona can be seen as a pinkish halo. The corona is the upper atmosphere of the Sun which extends outwards from the surface many millions of miles. Prominences are also visible during a solar eclipse as small reddish spikes or plumes of gas. With a spectrohelioscope the surface of the Sun can be observed, usually in

Fig. 25. Comet West, showing the features of the tail caused by solar wind.

the light emitted by hydrogen or calcium atoms. The instrument is tuned to one of these wavelengths and the resulting picture is one of a general background solar surface with bright and dark areas corresponding to emission or absorption by the atoms to which the instrument is tuned. For example if the atoms are emitting light at the wavelength to which the instrument is tuned, then these areas will appear brighter than the surrounding areas and if the atoms are absorbing light the areas will appear darker. This gives us an insight into the way that light reaches us from the Sun, and some of the processes of light emission at the surface. Solar flares can be observed with spectroheliscope. These are large clouds of gas which burst from the surface of the Sun and travel out into space. Although some of the material from the flares will return to the surface because of gravitational forces, a small percentage will escape from the Sun. We shall come on to the effects of this later.

This escape of gases from the surface of the Sun, together with a general leakage of particles, electrons, etc. causes streamers of gas to travel rapidly through the inner solar system. This is called the solar wind. The particles travel outward with a velocity of about 1 million miles an hour and the effects of this can be observed most readily in the comets. Take a close look at the photograph of Comet West (Fig. 25) and notice the twists or knots in the tail. These and the tail itself are caused by solar wind.

The Planets

MERCURY

Mercury is the first planet in order of distance from the Sun. Its equatorial diameter is 2970 miles (4780 km.), only slightly larger than our moon. It lies at a distance of only 35 million miles (57 million km.) from the Sun on average and so it never appears to be far away from it, 30° at maximum. Because Mercury lies so close to the Sun, it travels faster in its orbit than any other object in the solar system, its mean orbital velocity is 108,000 m.p.h. (174,000 km.p.h.), which means that even when the planet is visible it is only visible for a very short period. Mercury completes one sidereal orbit in 88 days approximately, and one synodic period in about 116 days. This means that Mercury can be seen at about the same place in the sky every 116 days.

Mercury is a very difficult object to observe, the two factors leading to this are its smallness, and hence the difficulty of being able to see any detail, and its close proximity to the Sun. Mercury shows a disc size of between 4·5 arc seconds at its smallest and 12·5 arc seconds at its best, so a small telescope will show a planetary disc, but even the largest telescopes will not show any surface detail. Some very important factors must be noted here relating to the finding of Mercury. Do wait until the Sun has set or try to find Mercury before sunrise. If the Sun is in the sky there is the chance that the telescope may be accidentally moved across the Sun with dire results. There will be no time to move your eye away from the eyepiece if you are not expecting a sudden burst of sunlight. Because Mercury's orbit lies inside the orbit of the Earth the planet will show phases similar to the phases of the Moon. Fig. 26 shows why this is so.

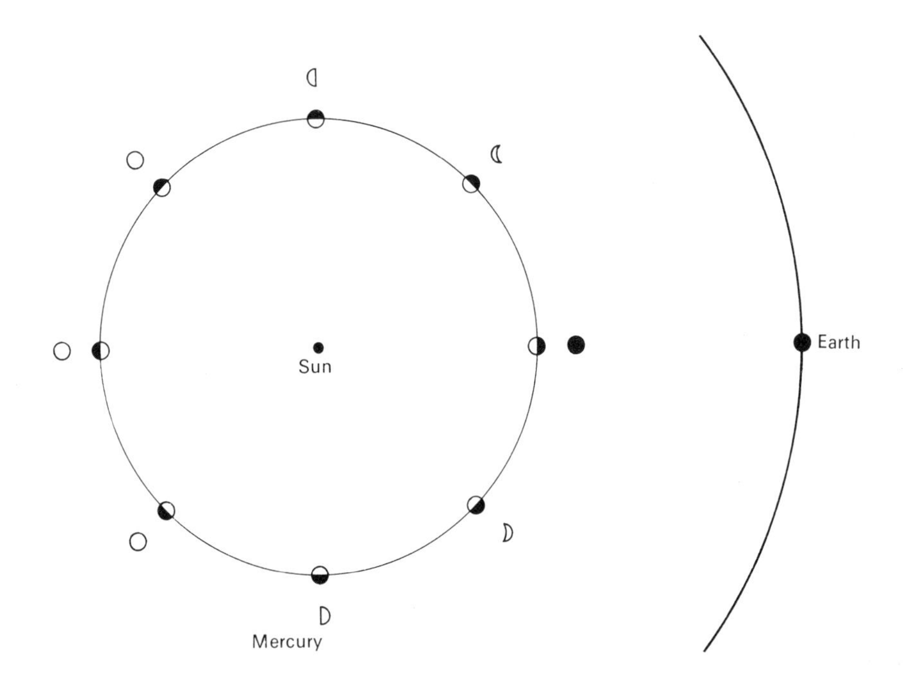

Fig. 26. Mercury and Venus show phases in a similar manner to our moon.

It was not until the advent of the space age that we really got to know the surface of Mercury. Expectation proved correct in assuming a crater surface, similar to the Moon. We knew that the surface would be very hot on the sunlit side and cold on the dark or night side. Present estimates of the temperatures at the surface are, for daytime 350 °C, which is about melting point of lead, and at night −230 °C. This large range in temperature is due to the fact that there is no atmosphere to speak of on the planet's surface to afford circulation of the heat, and Mercury's axial rotation period is about 58 days. This when coupled to the orbital period of some 88 days gives a full Mercurian day equal to 176 Earth days.

Visually Mercury is very much like the Moon, there are craters everywhere on the surface but unlike the Moon there are no maria or seas. The craters are almost certainly the result of impacts as there is little evidence to suggest that any of the craters have been produced by vulcanism. One interesting point arises from the pictures obtained by spacecraft that have visited the planet, and this is that the craters show a marked difference from the craters that we see on the Moon. Mercurian craters show a flatter structure than the lunar craters, often associated with concentric rings of material outside the main impact basin. These features are probably caused by the denser material from which the Mercurian surface is composed, as compared with the Moon's surface.

VENUS

Venus is the second planet in order of distance from the Sun. Venus has an equatorial diameter of 7700 miles (12,400 km.) and orbits the Sun with a sidereal period of 224 days. The synodic period of Venus is 583 days, which means that it is over $1\frac{1}{2}$ Earth years before it returns to roughly the same position in the sky. An interesting feature of this planet is that the axial period of rotation is 243 days, which is longer than the orbital period. This will effectively make the Sun appear, from the planet's surface, to move across the sky from west to east.

Venus must top the bill as the most inhospitable of the Earth-sized planets. The atmosphere is composed mainly of carbon dioxide with a small proportion of nitrogen. This is an extremely heavy mixture giving rise to a surface pressure of 100 Earth atmospheres at ground level. In addition to this the atmosphere contains a thick cloud layer which acts in the same way as the glass in a glass-house. Solar radiation reflected from the surface is trapped in the cloud layer causing general overheating in the lower levels and giving an incredibly high surface temperature of about 500 °C.

The clouds in the atmosphere of Venus contain, among other things, a high proportion of sulphur. The sulphur occurs as the element and also in combination with oxygen as the dioxide and trioxide, which combine with water to form sulphurous and sulphuric acids.

Observation of Venus is basically the same as of Mercury, but Venus lies more distant from the Sun than Mercury and is much easier and safer to observe. Venus can reach a maximum of 45° from the Sun and can be quite high in the sky before or after sunrise or sunset. The clouds on Venus, although obscuring the surface of the planet, reflect a high proportion of the sunlight falling on the uppermost layers. This, coupled with the fact that Venus can be quite close to the Earth, means that at favourable times Venus can be very bright. In fact Venus can be the brightest object in the sky apart from the Sun and Moon. The planet can reach magnitude −4·4 at maximum and has been known to cast a shadow under the best conditions. With a small telescope the phases of Venus are easily visible. In fact a fairly high power good quality binocular should show the phases, but I have yet to see this with a 16 × 50 binocular.

Several phenomena have been observed, such as the Ashen Light, which is a faint glow of light over the un-illuminated portion of the planet at about half phase. This is probably due to refraction in the upper atmosphere of the planet. Occasionally a slight blunting of the 'cusps' (the points of the planet at crescent phase) has been observed. Although there have been several suggestions as to possible causes no full explanation for this is available.

THE EARTH

We live on the planet Earth and are familiar with its immediate surroundings if not with all of the phenomena that are directly connected with this rather small insignificant object in space. The Earth is the third planet in order of distance from the Sun with an average distance of 93 million miles (150 million km.). This distance is recognised as a standard distance in astronomy called the Astronomical Unit or A.U., the value of which is taken to be 149,597,892·9 km. The Astronomical Unit is used for all 'in Solar System' measurement and is often incorporated into the measurement of stellar distances and the orbits of double or binary stars.

The Earth's equatorial diameter is approximately 8000 miles (12,000 km.) and it has an axial rotation period of 24 hours which gives a velocity of 1000 m.p.h. (1600 km.p.h.) at the equator. This may seem quite high at first, but we are travelling through space much faster than this, our orbital velocity is 70,000 m.p.h. (113,000 km.p.h.).

The Earth also has an atmosphere of course. This is composed mainly of nitrogen with about 20% of oxygen and traces of other gases of which argon is abundant, see Table 4.

Composition by volume	Height km.	Temperature	
100%H	1000		
75%H 15%He	700		Exosphere
		+	
70%N, 15%O, 15%He		2500 °C	
	500		
80%N, 15%O, 1% Ar, 1% Ozone		1500 °C	Ionosphere
	200		
		750°	
	100		
		−100 °C	
	50	0 °C	Stratosphere
	10	−50 °C	
78%N, 21%O, 1%Ar		18 °C	Troposphere

Table 4. The relative positions and popular names for the shells of atmosphere around the Earth with heights and temperatures. The Biosphere is the portion, restricted to part of the troposphere, the land and the seas, where biological activity, i.e. life, is found.

The density of the atmosphere decreases with increasing height above sea level and its composition also changes, the lighter elements of both helium and hydrogen becoming dominant. At 560 miles (900 km.) the composition is approximately 90% hydrogen and 10% helium. Initially from the surface of the Earth outwards, the temperature drops off rapidly, at 60 miles (100 km.) the temperature is about −100 °C but at 90 miles (150 km.) the temperature rises and by 250 miles (400 km.) is at 2000 °C. It must be remembered that in this context we are talking purely of temperature and not of heat capacity, the quantity of heat at this height is very small, see Table 4.

At lower layers in the atmosphere we find the clouds of water vapour. The Earth is the only planet in the Solar System which is known to have large quantities of free water available – 70% of the surface of the Earth is covered by the oceans. Also at lower altitudes we find life, from the simplest bacteria to the elephants. Life is certainly unique in the Solar System and if it exists elsewhere in the universe it must be very rare. Although there are many stars in our own galaxy around which life may have evolved, the margin within which a planet would have to be in order for life to exist and become intelligent is very narrow. If, for example, the radius of the Earth's orbit were more than 4% larger or smaller, it is almost certain that we would not be here today. Too close and the water on Earth would evaporate and cause a Venus glasshouse effect with runaway temperatures, too far away and the Earth's surface would be covered with glaciers. Something has caused the latter to happen in the not too distant past but at present the cause is unidentified.

Observable natural phenomena that are directly related to the Earth can reveal much information that would normally be inaccessible to an Earthbound scientist. The Earth's magnetic field is a good example. We can measure the field at the surface of the Earth but this reveals nothing of the field at altitude. The Earth receives all of its heat from the Sun. This energy supply arrives in a number of forms, such as light and heat radiation, and particles which are ejected from the Sun across the void to the Earth. Studies of the Sun and solar radiation and a particular Earth phenomenon revealed a striking correlation – namely the increase in frequency of Aurora sightings and the 11 year sunspot cycle. Then theory and observation combined to give the reason behind this correlation. Particles – electrons ejected during a period of increased solar activity – get tangled up in the Earth's magnetic field and directed towards the poles, both north and south, where they interact with atoms in the upper atmosphere of the Earth to form the characteristic glow of the northern or southern lights, the Aurora Borealis or Australis. Exactly the same process is occurring here that occurs when an electric current flows through a fluorescent light tube. Unfortunately the Auroral displays are only visible from very northerly or southerly latitudes and sightings from equatorial regions are virtually non-existent.

Other bodies that enter the Earth's atmosphere from space are the meteors and meteorites. The difference between the two is that meteorites are large objects which actually fall on to the Earth's surface. Meteors never fall to the surface in this way but burn up in the atmosphere, although the debris, very small particles of dust, may drift down and settle eventually but this takes a very long time. The meteors can be very useful objects to observe. They can reveal a great deal about the Earth's atmosphere and its density, and can provide hours of observing for the amateur. One of the popular tasks while observing meteors is to make a simple count of all the visible meteors during a given period of time. This can reveal details about the number of meteors in a particular shower, see Table 5, and the density of the meteors in their orbital stream.

Shower Name	R.A.	Dec.	Month	Max.
Coronae Australids	16·20	−48	Mar 14–18	16
Aquarids	22·24	0	May 1–8	5
Lyrids (June)	18·32	35	Jun 10–21	15
Capricornids	21·00	−15	Jul 10–Aug 15	Jul 25
Pisces Australids	22·40	−30	Jul 15–Aug 20	Jul 30
Perseids	03·04	58	Jul 25–Aug 18	Aug 12
Taurids	03·28	14	Oct 10–Dec 3	Nov 1
	03·36	21		
Leonids	10·08	22	Nov 15–19	17

Table 5. Details of major meteor showers with the month that they occur and the date of maximum activity.

The reason that we see a meteor as it enters the atmosphere is that the friction between the travelling object and the atoms of material in the atmosphere is so great that the temperature created melts the surface layers of the object causing a stream of material to flow behind it in a trail. It is this trail that we actually observe. If an object is large enough then there is a chance that not all of the material will be burnt away before the object is slowed down to the point where frictional forces are no longer great enough to cause continuous burning. Some of the material will then reach the surface of the Earth. This often happens in a dramatic way – the largest known crater on the Earth is one of about a mile in diameter. The mass which caused this has not yet been found and there is the possibility that as the object entered the Earth's surface the frictional forces were great enough to vaporise the object completely along with some of the surrounding area. The fragments that land are the meteorites. They are not all large but they are all useful. They can provide us with samples of material from other parts of the Solar System that would normally be inaccessible. The composition of meteorites can vary widely. Some are composed of iron or iron and nickel mixtures while others are stony with complicated compositions and many different structures. There are three basic classes of meteorite; the irons, stony irons and stones and these are further subdivided into as many classes as is necessary to incorporate the different types, possibly thirty or forty.

The last class of objects are the tektites. Their origin is not at all well defined and though many of these do show the ablation (burning up) effects of travelling through the Earth's atmosphere, there is no real evidence to suggest that they are definitely of extraterrestrial origin. The tektites are composed of an impure silica glass and their shape can vary from spheres and oblate spheroids to dumb-bell shapes. One of the peculiarities of the tektites is the limited distribution of examples over the surface of the Earth. They can be found in four regions; Texas and Georgia in the United States of America, the Ivory Coast of western Africa, Czechoslovakia in Europe and over parts of Australasia. Another peculiarity is that the tektites do not seem to be arriving at the present time and neither do they arrive over a long period of the Earth's history – the period for which we can calculate the arrival of the tektites is based on fossil dating of the soil type in which the tektites are found. This suggests that they arrived on the Earth between 30 and 10 million years ago.

THE MOON

The Moon is the Earth's natural satellite, something possessed by

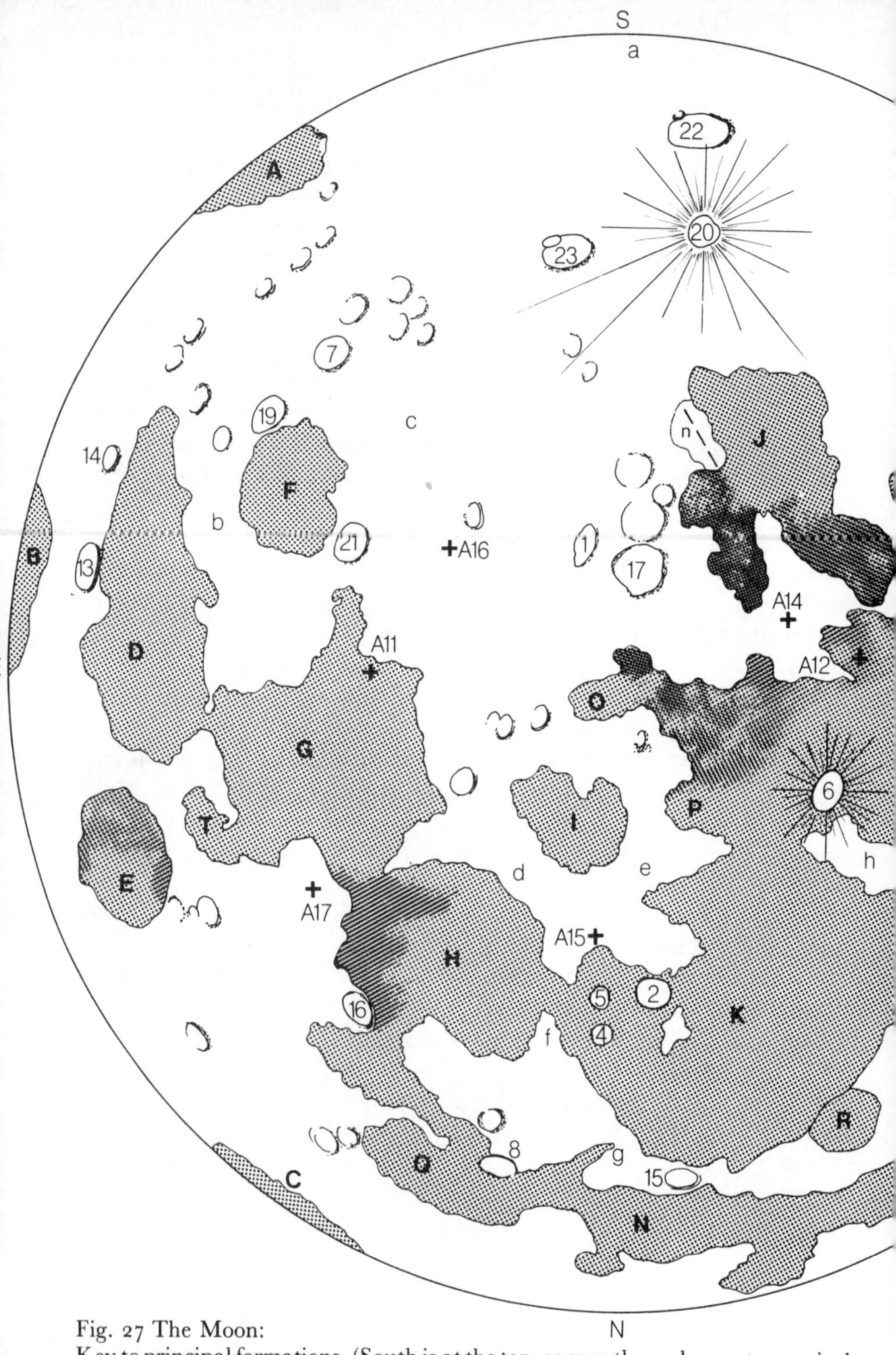

Fig. 27 The Moon:
Key to principal formations. (South is at the top, as seen through an astronomical telescope. Seen through binoculars, the image would be the other way up, with south at the bottom.)

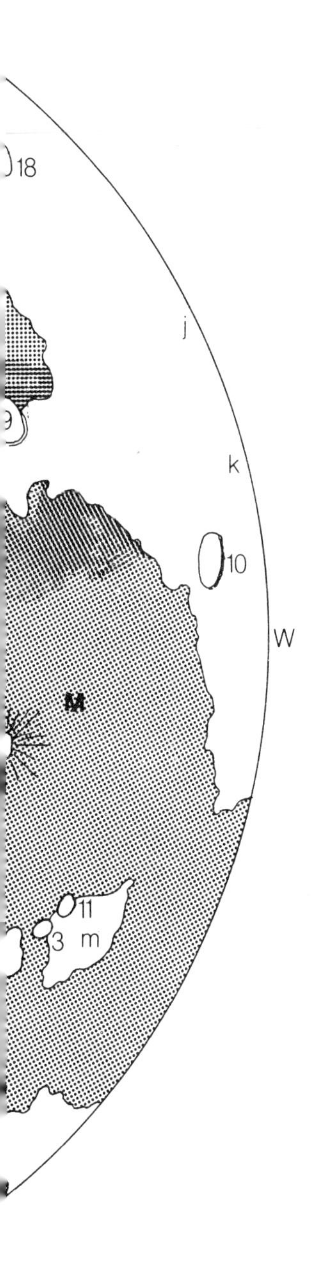

Craters:
1 Albategnius
2 Archimedes
3 Aristarchus
4 Aristillus
5 Autolycus
6 Copernicus
7 Piccolomini
8 Aristoteles
9 Gassendi
10 Grimaldi
11 Herodotus
12 Kepler
13 Langrenus
14 Vendelinus
15 Plato
16 Posidonius
17 Ptolemaeus
18 Schickard
19 Fracastorius
20 Tycho
21 Theophilus
22 Clavius
23 Stofler

Mountain Features:
a Leibnitz Mts
b Pyrenees Mts
c Altai Mts
d Haemus Mts
e Apennine Mts
f Caucasus Mts
g Alpine Valley
h Carpathian Mts
i Riphaen Mts
j Cordillera Mts
k D'Alembert Mts
m Schröter's Valley
n Straight Wall

Mare Features:
A Mare Australe
B Mare Smyth
C Mare Humboldtianum
D Mare Foecunditatis
E Mare Crisium
F Mare Nectaris
G Mare Tranquillitatis
H Mare Serenitatis
I Mare Vaporum
J Mare Nubium
K Mare Imbrium
L Mare Humorum
M Oceanus Procellarum
N Mare Frigoris
O Sinus Medii
P Sinus Aestuum
Q Lacus Somniorum
R Sinus Iridum
S Sinus Roris
T Falus Somnii

Apollo manned landing sites
A11 1969
A12 1969
A14 1971
A15 1971
A16 1972
A17 1972

neither Mercury nor Venus. Although we regard the Moon as quite a small body, when the ratio of the sizes of the other primary planets to their satellites is considered, it is almost as if we are living on the primary object of a double planet system rather than a planet-moon system. The Moon is large when compared with the size of the Earth and other primary-satellite systems. The Earth is some eighty-one times more massive than the Moon, but Saturn is 4100 times more massive than its satellite Titan, the largest satellite in the Solar System. The Moon lies at an average distance of 249,000 miles (384,000 km.) from the Earth and is about 2000 miles (3,476 km.) in diameter extending across some 31 minutes of arc of the sky.

The Moon is probably the first object that the beginner will want to look at through a telescope – and what an object it is! See Fig. 27.

There are several features on the surface of the Moon which are interesting to look at. Even a small telescope will show the large dark areas called maria, or seas, and some of the larger craters. Basically the mare are flat plains caused by basaltic lava flows into crater basins. After an impacting object has collided with the Moon leaving a large crater, the semi-liquid lava slowly seeps out of the damaged crust to form a lake in the impact basin. Smaller impacting objects do not have the energy to break the crust sufficiently to cause this flow of lava and these craters remain as they were formed. Subsequently some areas of the Moon are covered with small craters, see Fig. 28. This is a photograph of the area around the crater Tycho which is 53 miles (85 km.) in diameter and is one of the more recent features on the Moon's surface. The object which formed the crater Tycho also caused the ray structure associated with this crater. The rays can be seen quite easily with a small telescope or binocular and extend across the visible hemisphere of the Moon for a distance almost 1800 miles (3000 km.) Ray structures are caused by material ejected by the impacting object as it hits the surface at an extremely high velocity. They are not restricted to the crater Tycho – many of the more recent craters have rays but Tycho's are by far the most prominent.

Other features that can be observed on the surface of the Moon, such as the rilles and faults, really need a fairly large telescope to reveal them. The rilles are collapsed lava tubes, rather like collapsed underground caves on the Earth such as Cheddar Gorge in England. Faults on the Moon, though not as common as the rilles,

Fig. 28. The area surrounding the crater Tycho on the surface of the Moon, photographed from the Tattershall Bridge Observatory.

are much more striking. The best time to find a good example of a fault is just after the Moon has reached its half phase, when the Straight Wall is easily visible with quite small instruments. Valleys on the moon are quite common, but are not easy to see. Very often so-called valleys are formed by successive cratering in one area, the adjacent crater walls forming the sides of the valley. One of the largest valleys on the Moon is Schröter's Valley.

Another feature which has become more eminent recently are the T.L.P.s or Transient Lunar Phenomena. T.L.P.s are fairly short-lived events during which time a bright crater such as Tycho becomes obscured by the emission of gas from somewhere within it. Present theory suggests that the periodic escape of gas from below the surface, rather like a geyser issuing steam, is the cause of this but the reason for this emission is still unknown. T.L.P.s are observed as pinkish clouds or simply a general obscuration of any surface detail around the immediate area surrounding the crater. Observation of

T.L.P.s is carried out with the aid of colour filters in the light path of the telescope. This usually consists of some method for holding a red and a blue filter in such a way that they can be alternately brought into the light path. Any alteration in colour will show up as a difference in contrast between the images viewed through the two filters, see Fig. 29.

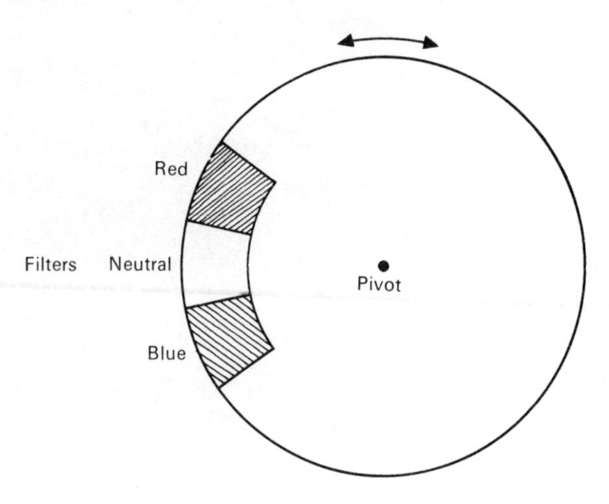

Fig. 29. A colour filter placed behind the eyepiece of a telescope to reveal contrast differences on the surface of the moon.

In Fig. 29 the disc is placed in the light path of the telescope so that the image is passed through the filters; by rocking the filters back and forth the type of filter can be changed quite rapidly. It is usually found that the frequency of oscillation has to be about twice per second to detect changes on the surface.

Lunar photography is an easy project for the beginner. It requires the use of preferably a single lens Reflex camera so that the image to be photographed can be observed up to the instant of exposure. The film should be of fairly fine grain and exposure times will vary with film type, telescope type and magnification, so there is plenty of room for trial and error methods with any particular set-up. It is of course essential that copious notes are made of the method by which each exposure is made so that if a good result is obtained it can be repeated. To take a photograph of the Moon, first remove the camera lens, if this is possible, and using a low-powered eyepiece with the camera placed close to it focus the image by means of the normal eyepiece draw tube or rack focus until it is as sharp as possible. Then gently squeeze the shutter release keeping as still as possible to avoid vibration. Using a shutter speed of about 1/500 second to 1/250 second for a half-Moon. This will not give excellent results but it will give an image on the film which should reveal the mare and brighter craters. To refine this technique it becomes necessary to use a camera mount on the telescope and an electric drive unit to keep the image steady in the eyepiece. Many specialist books are available that deal with these and similar associated topics.

MARS

Mars is the fourth planet in order of distance from the Sun. It lies at an average of 1·5 A.U. or 141 million miles (228 million km.) approximately. The equatorial diameter of this planet is only 4220 miles (12,390 km.) and thus it extends an image of only 18 seconds of arc to the Earthbound observer – at its worst this can be as small as 3·5 seconds of arc. The sidereal period of Mars is 687 days and this coupled with the Earth's orbital period gives Mars a synodic period of 780 days or just over two Earth years. This means that Mars only reaches the same position relative to the Earth and Sun once in every 2 years. Therefore when Mars is in the sky, make full use of it because it will disappear for quite some time behind the Sun.

As a planet Mars is reasonably hospitable. There is very little atmosphere and this is composed chiefly of carbon dioxide with a little nitrogen and argon and about 0·1–0·3% oxygen. The surface temperature varies widely from a maximum of 16 °C to a minimum of −130 °C and this variation in temperature can cause large pressure differences leading to high surface wind speeds of up to 300 m.p.h. (500 km.p.h.). Because there is very little atmosphere though, the wind is not very energetic and speeds of 100 m.p.h. (160 km.p.h.) are required just to move the lightest dust particles around on the surface. The surface air pressure of Mars's atmosphere is only about one-tenth that of the Earth's.

The surface of Mars is rocky in places with sand dunes. The surface soils have a red colour which is quite easily visible from the distance of the Earth. There are some impact craters on the surface of Mars but these are dwarfed by the massive shield volcanoes in the Tharsis region, the largest of which, Nix Olympica (or Olympus Mons), is 17 miles (27 km.) in height. The Coprates Canal is a canyon system which makes the Grand Canyon on Earth look rather insignificant. The Coprates system extends for some 1500 miles (2500 km.), 100 miles (160 km.) in width and has a maximum depth of about 5 miles (8 km.)

Conditions of observation of the planet can vary and are largely dependent on the position of both Earth and Mars in their respective orbits. When Mars is at perihelion – closest point to the Sun – and the Earth is at aphelion – farthest point from the Sun – and when Mars is at opposition and favourable for observation, even small telescopes will show the planetary disc and possibly some dark and light patches on the surface. The angular diameter of Mars at this time will be 25·6 seconds of arc. But even so the markings on this planet are very delicate and very large telescopes are needed to show good surface detail. To carry out useful work with Mars really requires a 6 in. (15 cm.) refractor or 8–10 in. (20–25 cm.) reflector, however don't let this put you off having a look through even a small instrument at the planet.

Mars like the Earth has polar caps. On Mars they are composed of mainly dry ice – solid carbon dioxide – with some water ice. Under good conditions a 3 in. (7 cm.) refractor will show these to be white in contrast to the reddish surface of the planet.

During the Martian summer the caps are quite small, only covering a few hundred square miles of surface area, but as the winter progresses the caps extend over a larger surface of the planet towards the equatorial regions. Martian dust storms can also be observed as winds rush around the planet at speeds of 300 m.p.h. (500 km.p.h.), transporting dust particles across the surface and creating the sand dunes that have been observed by the Viking spacecraft. The dust storms are usually severe enough to obscure large areas of the planet as seen from the Earth and useful observation can be carried out here with a medium-sized telescope by monitoring the dust storms as they appear. Useful work can also be carried out by monitoring the procession and recession of the polar ice caps.

Mars has two moons, Phobos and Deimos. These two small

chunks of rock, Phobos 7·5 miles (12 km.) in diameter and Deimos 5 miles (8 km.) in diameter, orbit Mars at distances of 5700 and 14,500 miles (9200 and 23,500 km.) respectively, are often given the nickname of the 'potatoes' owing to their shape and size. They are in all probability captured asteroids. The brighter of the two, Phobos, shines at the 11th magnitude and Deimos about the 13th magnitude.

THE ASTEROID BELT

In the last section we referred to Mars's moons as possibly being captured asteroids. The word 'asteroid' literally means 'star shaped' or 'star like'. The first suggestion of objects that we now call the asteroids came during the late eighteenth century when the mathematician J. E. Bode calculated that there was a planet missing between the orbits of Mars and Jupiter. This formula used by Bode, now called Bode's law, was a quite simple mathematical relationship between the planetary distances. Take the numbers 0, 3, 6, 12, 24, etc. (each of which is double the previous figure) and add the figure 4 to each. By dividing the resultant numbers by 10, an approximation to the distances in A.U. of the planets can be found, see Table 6. It was 1801 when the search for the missing planet produced the result of finding a small faint object of about the 7th magnitude. Calculation of the orbit of this new member of the Solar System agreed very well with the prediction as calculated from Bode's Law. The object became named Ceres and had an orbit with a mean distance of 2·76 A.U. The search continued and revealed three more of these objects, now called Pallas, Juno and Vesta.

Planet	Bode's Distance A.U.	Real Distance A.U.
Mercury	0·4	0.38
Venus	0·7	0·72
Earth	1·0	1·0
Mars	1·6	1·52
-------- (Asteroids)	2·8	----- (2·8)
Jupiter	5·2	5·2
Saturn	10·0	9·54
Uranus	19·6	19·18
Neptune	38·8	30·0
Pluto	77·2	39·44

Table 6. Bode's Law in tabulated form to show the close corellation of the law to the planetary distances. Distances are in Astronomical Units.

Asteroid Name	Dia. Miles	Km.	Mean distance from Sun A.U.	Inc.° Orbital	Period years	Mag. vis.
Ceres	425	685	2·76	10·5	4·6	7·4
Pallas	279	450	2·76	34·8	4·6	8·9
Juno	149	240	2·66	13·0	4·3	8·0
Vesta	336	590	2·35	7·1	3·6	6·0

Table 7. The four brightest Asteroids.

From the mid-nineteenth century the list of known asteroids continued to grow and at present accurate orbits for about 2000 objects are known. Many more have been seen either directly or by long exposure photographic methods, but have been subsequently lost. Present estimates suggest that there are more than 40,000 of these bodies with a diameter of more than a mile with numbers increasing with decreasing size.

The four brightest asteroids mentioned above can all be seen with a small telescope or binocular and Vesta can, under good conditions, be seen with the naked eye. To find an asteroid, reference should be made to an almanac or suitable year book.

Not all of the asteroids have orbits which lie within the Bode suggested mean. Some have orbits which are very eccentric and the asteroids may approach closer to the Sun than does the Earth, whilst their aphelia may reach the orbit of Saturn in extreme cases. One group of asteroids – the Trojan Asteroids – occupy what are known as the Trojan points. These are positions at an equal distance from the Sun and Jupiter and in the orbital plane of Jupiter where gravitational forces are focused and an object can be trapped until perturbed by another object into a different orbit.

JUPITER

The giant planet lies at a distance of 5·202 A.U. from the Sun, has an equatorial diameter of 88,700 miles (143,000 km.) which is about 10 times the diameter of Earth, and is 300 times more massive than Earth, with a volume 1300 times that of the Earth. The composition of Jupiter is chiefly hydrogen and helium, with hydrogen probably accounting for about 80% of the mass. Other elements are also observed to be present on this planet but they are in very small amounts. Ammonia, methane and water have all been observed spectroscopically from Earth. We cannot discuss the surface features of this planet because there is no real surface. The atmospheric gases gradually increase in density to form a liquid semi-surface. This continues through the planet until the pressure and density and temperature are great enough to convert the liquid hydrogen to the solid metallic phase. Jupiter possibly has a very small rocky core of only a few thousand miles in diameter.

Jupiter rotates on its axis once in 10 hours approximately and with a circumference of 287,000 miles (460,000 km.) the surface is speeding round at 28,700 m.p.h. (46,000 km.p.h.) which has a terrific torsion effect tending to spin the outer layers at the equator outwards by centrifugal forces. Even a small telescope will show that the equatorial diameter is greater than the polar diameter. The amount of difference between these is about 6000 miles (9700 km.) As on the surface of the Sun, the equatorial regions of Jupiter are rotating much faster than the polar regions, causing rotation belts and zones of different colours on the visible surface. The colours are

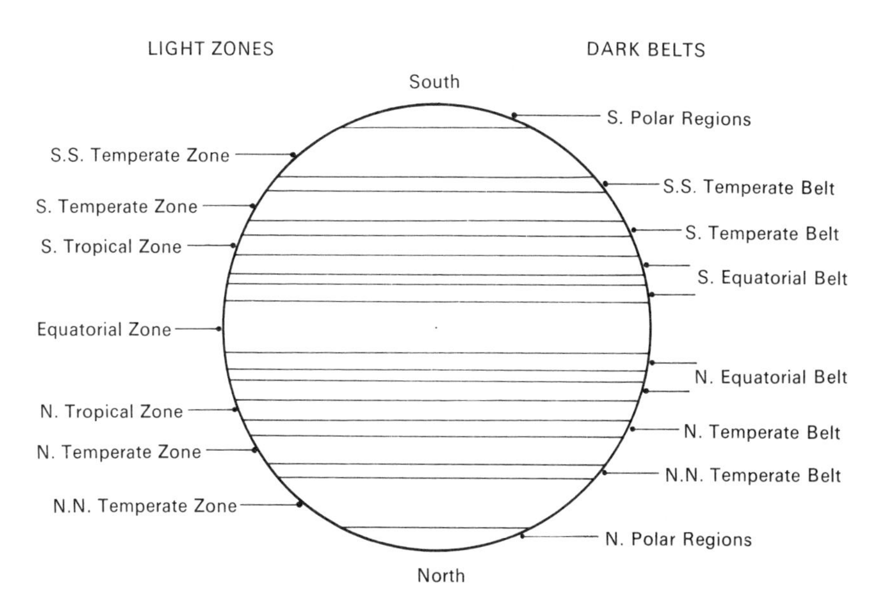

Fig. 30. Zones and belts on the surface of the planet Jupiter. Observers looking for the Red Spot should find it in the S. Tropical Zone. It is customary to place south at the top of such a diagram, as this is the way the image is presented to the observer through an astronomical telescope.

due to the convection pattern, warmer gas rising to the surface in some areas and cold gases sinking back to the lower layers in others, see Fig. 30.

This gives rise to some interesting phenomena. Frictional forces in the upper layers cause a massive buildup of static electricity which is released in the form of giant electrical storms over the surface. This may help to form some more complex molecules from the mixture of gases already there giving rise to the coloration which is observed. Colours range from off-white through yellow and orange to dark reddish-brown.

The temperature of Jupiter at the surface is quite cold, about $-150\,°C$. One of the mysteries concerning Jupiter is that the planet seems to be radiating more energy from its surface than it is receiving from the Sun. Several theories have been put forward to explain this and one such theory suggested that the planet, like the Sun, is massive enough to support thermo-nuclear reactions.

This has now been virtually ruled out and the present explanation assumes that all of the planets formed from the solar nebula by contraction, the solar nebula being a large cloud of dust and gases, rather like the Orion Nebula. The gravitational attraction of the particles within the cloud caused a slow contraction of the whole cloud to take place. The central condensation formed the sun whilst smaller wells of material collected at various points and became the present day planets. During the contraction process heat was generated and retained in the sun and planets. The sun became hot enough to start a thermo-nuclear reaction but the planets, over the years, have lost much of the original heat. Jupiter, though, is large enough to have retained some of the heat and this is the radiation that is observed. The giant planet Jupiter is still contracting from the original formation of the solar system, and consequently cooling down.

Jupiter is probably one of the objects in the solar system most frequently observed by amateurs. It presents an image which even the smallest telescope will show as a disc, whilst a moderate sized objective will show considerable detail in the cloud belts themselves. A 3 in. (7 cm.) refractor telescope will show the polar flattening, and another feature in the cloud belt south of the equator called the Great Red Spot. A huge swirling cloud of gases, similar to a terrestrial tornado but much larger and more persistent, the great

Red Spot has been observed since the time that a large enough telescope was built to observe it. Occasionally the Spot has almost disappeared and at other times it is very prominent indeed showing as its name suggests, a deep red colour. But in the long run very little change has been observed in this feature of Jupiter. Short-term variation in the clouds are among the popular observing topics. The visible surface of Jupiter can vary considerably in a short space of time, but this usually requires the use of a fairly large telescope. Delicate changes in the clouds take place, usually due to disturbances between the layers or often in association with features such as the Red Spot.

Jupiter has a great many moons, at least thirteen have been observed and there are possibly more smaller objects associated with this planet. Because Jupiter is so massive the gravitational attraction towards close objects is immense. Asteroids in the vicinity of Jupiter can be easily trapped and made to orbit the planet at great distances. The four major moons of Jupiter are very famous objects indeed. They were first observed by Galileo and are hence called the Galilean moons. Later they became the objects which first helped to measure the speed of light. At periods of one Earth year, the periods of the moons are observed to lengthen and shorten by 1000 seconds. It was suggested that this may be due to the fact that as the Earth moves in its orbit, the distance between Jupiter and the Earth changes by approximately the distance across the Earth's orbit – 186 million miles (300 million km.). Dividing this by the difference in the orbital periods of the Jovian satellites gives a rough approximation to the velocity at which light travels across the Earth's orbit – approximately 186,000 miles (300,000 km.) per second. The first four large moons, called Io, Europa, Callisto and Ganymede, are at the present time (April 1979) undergoing scrutiny by two spacecraft along with the planet Jupiter itself. The results from the moons are more surprising than those from the planet. The moon Io has been observed at fairly close range and on Io are the only known active volcanoes in the solar system apart from those on Earth. A giant crater has been observed on the moon Callisto, several hundred km. across. One important point which relates neither to the planet itself nor the moons is that Jupiter has a ring system. At present no firm details are known regarding its size or thickness. These are just some of the interesting features that modern technology and spaceflight are bringing to our notice. For more detail of the physical and orbital details of Jupiter's four bright moons refer to Table 8.

Name	Mean distance from Jupiter Miles	Km.	Period	Mag. (vis.)	Diam. Miles	Km.
Io	262,000	422,000	1·769	4·7	2274	3659
Europa	417,000	671,000	3·551	5·2	1800	2900
Ganymede	665,000	1,070,000	7·166	4·6	3100	5000
Callisto	1,170,000	1,882,000	16·753	5·6	2870	4620

Table 8. Data for Jupiter's major satellites. These can be seen with a low-power binocular.

SATURN

Saturn, like Jupiter is a large gaseous body with an equatorial diameter of 74,000 miles (119,000 km.) The polar flattening of Saturn is rather more evident than that of Jupiter. The polar diameter of Saturn is only 66,900 miles (108,000 km.) Saturn takes 29 4 years to make one orbit around the Sun at a mean distance of 9 54 A.U. or 886 million miles (1143 million km.) The appearance of Saturn and its chemical composition are both very similar to those of the planet Jupiter, hydrogen and helium being the main constituents giving rise to a very similar surface. We can observe spots on the surface of Saturn in the equatorial regions, though they are not in any way as prominent as the features on Jupiter. The surface temperature of Saturn is about −170 C.

Visually Saturn is one of the most beautiful objects in the sky. Through a binocular the planet appears to be elliptical and with larger instruments, the elliptical shape is resolved into a fantastic system of rings. Although a large telescope is needed to reveal the delicate structure in the rings, a 3 in. (7 cm.) refractor should under reasonable conditions show the main ring system. There are in fact three rings in the system, usually given the designation A, B and C in order of decreasing radius. Ring A is rather faint and in small telescopes this merges gradually into ring B but in larger instruments there is a distinct gap between the outer two rings. This is the Cassini division, named after G. D. Cassini who first recognised the gap. Ring C is often called the crepe ring, it is quite faint and needs a fairly large telescope to reveal it. At times observers have often filed the claim that they have seen another ring even closer to the planet, but this has not as yet been verified. From year to year, due to the tilt of the axis of the planet Saturn, the ring system can be seen with various inclinations to the Earth. At times the rings will be wide open and at others edge on and invisible, except with the largest telescopes. The tilt of Saturn's axis to its orbital path is 26°.

Saturn has a large family of satellites. At least ten are known, the smallest of which is only a little over 100 miles (160 km.) in diameter, and they may not be all of the family. It is quite possible that we have only observed a small number of the satellites and that there may be a great swarm of them just waiting for the Voyager spacecraft. The brightest of the satellites, Titan at magnitude 8·4, has an equatorial diameter of 5000 miles (8000 km.) This moon may just be seen with a small telescope.

URANUS

This planet lies at a distance of 19·18 A.U. almost 1800 million miles (2900 million km.) and has an equatorial diameter of 32,000 miles (48,000 km.) Whilst this may seem quite large at first the planet only extends an image of some 4 seconds of arc which is very small in any telescope. From Earth even the largest telescopes will not show much detail. The planet appears blue in colour and also shows some striations in the upper layers which are due to a rapid rotation period of only 10 hours or so (10·8). The planet can be found with a binocular and at its best can be seen with the naked eye. It shines at magnitude 5·8.

One of the ways to determine more about a planet is to calculate possible times when the planet is likely to occult or move in front of a star. Because the light from a star is effectively a point source, it is possible to observe the star as it passes behind the atmosphere of the planet and eventually behind the planet's surface. It was during one of these exercises that it was noticed that the light of the star was dimmed considerably at a distance from the planet which suggested that there may be a ring system similar to that of Saturn. The ring is thought to be about 31,000 miles (49,000 km.) in diameter and latest observations suggest that the ring is horse-shoe shaped. The ring is composed of small rock particles covered in ice, the size of particles ranging from a little greater than 100 miles (160 km.) to small dust grains. Uranus has five satellites, the brightest of which is of the 14th magnitude and requires a 12 in. (30 cm.) telescope to see them.

NEPTUNE

The eighth planet in order of distance from the Sun, this planet remains beyond naked eye visibility to the Earthbound observer. It lies at a mean distance from the Sun of 30 A.U., 2800 million miles (4500 million km.) and with an equatorial diameter of only 30,000 miles (48,000 km.) presents an image of 2 seconds of arc across. It orbits the Sun in 165 years and only changes its relative position in the sky very slowly. The synodic period of Neptune is 367 days.

Neptune has two satellites, Triton and Nereid. Triton is 3700 miles (6,000 km.) in diameter and at this distance shines at magnitude 13·5 and can be seen with an 8 in. (20 cm.) or 10 in. (25 cm.) telescope.

PLUTO

This unusual little creature on the outer edges of the solar system reveals very little about itself. Pluto is small and a long, long way away. At its closest it is slightly nearer to the Sun than is Neptune, but at its farthest it is 3600 million miles (5900 million km.). As even the largest telescopes will not show Pluto as a disk, no detail can be seen on the planet and our only hope is the spectroscope, but even this has failed to reveal any hard facts about a possible atmosphere on this planet.

Pluto is thought to be about 4000 miles (5900 km.) in diameter and has a mass of about a tenth of the Earth's. It takes 247 years to make one sidereal orbit around the Sun and rotates on its axis in 6·4 days. Because this planet is so small and lies at such an immense distance it is very faint, at its best can only reach about magnitude 14.

The Comets

The comets are small spheres of gas and dust that by the nature of their highly elliptical orbits only approach the Sun once in a great number of years. Some comets are thought to have periods of several thousand years.

As a comet approaches the Sun some of the gaseous material is driven off. The pressure of the solar wind then pushes this material out in the form of a tail stretching out behind the head or nucleus of the comet, always trailing away from the Sun.

The origin of the comets is a little ill defined as yet but they are possibly a mixture of small rock particles left over from the formation of the planets together with the gaseous component ejected from the Sun during the T Tauri stage of the Sun's early evolution. The T Tauri stage was the stage immediately after the contraction from the solar nebula, as the hydrogen atoms begin to combine to form helium. The release of energy at this stage was enormous and almost explosive. The outer layers of a star at this stage of evolution are pushed outwards to accommodate the increase in pressure at the centre. Such a star is variable in nature in its light output until an equilibrium is reached when the energy released from the centre of the star is balanced by the gravitational contraction. The name T Tauri comes from the variable star T in the constellation of Taurus, which is at this stage of evolution at the present time. Present theory suggests that a large proportion of stellar material would be ejected outwards from the Sun during its early life and this material could condense to the small globules that we now call the comets.

The comets seem to originate from a position some 60,000 A.U. from the Sun, by tracing the orbits backward. It would further be quite feasible to assume that there are a large number of comets at approximately this distance in stable, almost circular, orbits. Any periodic planetary configuration such as the alignment of four major gaseous planets every so many years could gradually perturb the orbits of comets sufficiently to take them into elliptical, parabolic or, in an extreme case, hyperbolic orbits bringing them into the close perihelion passages that we now observe.

The comets, especially the large long period comets, are spectacular. The tails of these can reach giant proportions visible across many degrees of sky, but these are seldom seen. Only a handful of the really spectacular comets have been observed on more than one return. One of the most noted is of course Halley's comet, which has been observed for many hundreds of years on successive returns to perihelion. Halley's comet is due to return again around Christmas

1985 and will be easily visible to southern hemisphere observers.

Visually the comets appear as small diffuse misty patches of light at the limit of the telescope's power, or very rarely as the large bright nucleus with a tail which stretches right across the sky. Hunting for comets can be carried out quite simply with modest equipment such as a binocular or more usually a wide field telescope. Because the comets are so rare, many hours can be spent on hunting them without any results. The best place to search is near to the Sun just before sunrise or after sunset. Sweeping the area thoroughly and carefully for anything that may look like a comet. The reason for searching near the Sun is that this is the point where the comet will be at its brightest. If a suspect is found it should be checked against the star atlas to make sure that there is no ambiguity between the object and a star cluster or galaxy or any other marked object. Make a careful note of the position of the object and the stars in the field of view and near the comet. After about an hour check the position again and if the object is a comet, the relative position of the comet and nearby stars will have changed. Because of the close proximity of a comet to the Sun, the comet will move through the background quite rapidly and a significant change will be noticed over a period of about an hour.

Earlier in this chapter we discussed the boundaries of the Solar System. It will be noticed that there is a large anomaly in the two parameters, the solar gravisphere and the cometary orbital distances. With present knowledge it is not possible to explain this anomally fully. We await further observational data.

6

The Galaxies

All of the objects that we observe from the Earth, or for that matter anywhere, are parts of giant systems called galaxies. These are huge rotating groups of stars, containing many hundreds of millions. Not many observers will have the use of a telescope that can show the external galactic systems and so for the major part of this chapter we shall be dealing with our own galaxy, the galaxy in which the Sun is situated. The first indication of the structure of the galaxies came when Lord Rosse made the famous drawing of M51 in Canes Venatici in 1845 and suggested that the spiral structure may be caused by the rotation of a multitude of stars within the structure. This has been proved to be correct, the Sun makes one revolution about our galaxy in 250 million years.

Present day calculations show that there are roughly 100 thousand million stars in our galaxy. Some of the larger galaxies contain many more than this, but for most purposes our galaxy can be taken to be a good average size.

At a first look around the sky, it may seem as if the stars are rather haphazardly strewn around the galaxy but this is not at all the case. The stars in each galaxy are arranged in a variety of definite patterns depending on the galaxy type. Galactic classification is to be our first topic in this chapter. Although there is no distinct evolutionary process by which we can catalogue the different types of galaxy, there are a limited number of shapes that we can recognise.

There are four basic types of galaxy with sub-groups in each type. These are spiral, barred-spiral, elliptical and irregular.

The spirals are subdivided into three categories; Sa, Sb and Sc galaxies. An Sa type galaxy has a large well defined nucleus and tightly wound spiral arms with a great deal of gaseous material present. These galaxies have a high surface brightness, probably due to the gas and dust in the arms. The arms in a Sa type galaxy are often so closely wound as to be indistinguishable from the nucleus and thus are closely related to the SO or lenticular galaxy which we will come to later. The Sb types have a large nucleus, but the arms tend to be much more open and well defined. There is still a large amount of gas and dust in both the nucleus and spiral arms, but not to the extent of an Sa galaxy. Our own galaxy is an Sb type. Sc type galaxies have a very low surface brightness and the arms are diffuse and indistinct with little or no gas in them, though the nucleus may show bright knots of nebulosity.

The barred spiral galaxies are a variation of the normal spiral types and, as the name suggests, have a bar extending through the nucleus joining the arms. The barred spirals are classified in much the same way as the ordinary spirals, there being three classes, SBa, SBb and SBc. An SBa galaxy has closely wound arms often almost reaching the bar of the second arm to form a complete circle with the bar across the diameter. Much dust and gas is evident along the bar and in the arms. An SBb galaxy, whilst retaining the gas and dust in the bar and to some extent in the arms themselves, has shorter arms with less material. The SBc galaxies have a prominent bar, rich in gas and dust, but the arms become short extensions of this and are very often ill defined. The SBc galaxies may be difficult to distinguish from irregular types.

The irregular galaxies have, as the name suggests, little or no structure, and as a general rule gas and dust is very limited in quantity. The stars are highly evolved and have used up most of the gaseous material.

The elliptical galaxies have a very regular form and are placed in six groups of increasing ellipticity from E0 to E5. For some reason the elliptical galaxies show highly evolved features. No gas or dust is visible in these galaxies and the stars show the features of great age.

A further class of galaxy, the lenticular type, can be likened to two dinner plates placed face to face. They are in some ways like a spiral galaxy without arms. Very little internal structure can be seen in a lenticular galaxy but they often have a visible dust lane in the equatorial regions. The lenticular galaxy is given the designation of S0.

For a schematic view of the classification of the galaxies see Fig. 31.

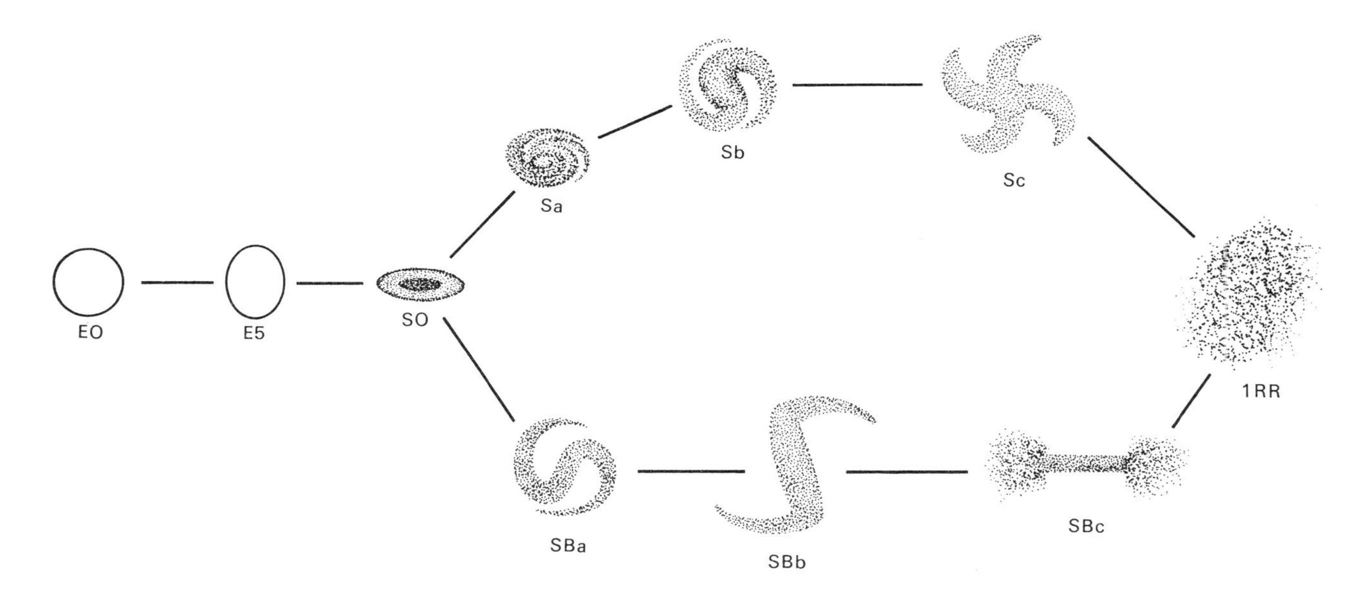

Fig. 31. This representation of our system of classification of galaxies by no means relates to a schematic view of the evolution of a typical galaxy, but rather presents an easy to look at way of explaining the different types of galaxy that we can observe from the Earth.

As already mentioned our own galaxy is an Sb type. Our position in the galaxy is slightly away from the plane and about two thirds of the distance from the nucleus to the outer edge, see Fig. 18. The nucleus of the galaxy as seen from our position lies in the direction of Sagittarius, a southern constellation. The milky way is actually the plane of the galaxy and this can be seen quite easily on a clear moonless night as a band of misty light stretching across the sky. The number density, or number of stars per unit volume, in the milky way is considerably higher than it is at a short distance from the plane, for example at positions X and Z in Fig. 18. A binocular, with its wide field of view, will show this remarkably well. The plane of the galaxy is the most highly populated region and numbers of stars increase rapidly towards the nucleus. The other constituents such as the gas and dust clouds already mentioned are also denser in the plane. Southern hemisphere observers should sweep through the constellation of Crux, which can be found on the star atlas, where a large dark patch can be seen. This patch is a huge cloud of obscuring dust that is stopping the light from stars behind it from reaching us. Other clouds have stars embedded in them, causing the gas to glow. These form the well-known gaseous nebula such as Eta Carina (Plate 1) and the Orion Nebula (Plate 6).

Also confined to the plane of the milky way are the star clusters. The open or galactic clusters are large clouds of stars which are in gravitational contact with each other and hence they remain intact for much of their lives. Only effects due to the rotation of the galaxy and closely passing stars tend to cause the clusters to dissociate, and this takes a long while. The galaxy rotates once in 250 million years, and the Sun, at 4000 million years old, has completed about sixteen rotations. Over a long period time the clusters may become loose and contain fewer members than the younger clusters. Good examples of these clusters are Herschell's Jewel Box in the constellation of Crux, the Pleiades in Taurus and M44 the Beehive in Cancer. A search of the plane of the galaxy will reveal more of these interesting objects but there are some clusters in the galactic plane which are difficult to observe and were not discovered until the spectroscope came into wide use on the larger telescope systems. These clusters are composed of stars of a particular spectral class such as the O and B types – hot blueish stars thought to be of very recent origin. Of even more recent formation are the T associations

1 Eta Carina is one of the brighter gaseous nubulae in the Southern Hemisphere.

2 The Large Magellanic Cloud is a small satellite galaxy to our own at a distance of about 170,000 L.Y. in the constellation of Dorado.

3 The Rosette Nebula was probably formed after a recent supernova and the gases have retained a roughly circular shape.

4 The Lagoon Nebula, a fairly bright example of nebulosity in Sagittarius.

5 The Trifid Nebula is a small gas cloud and, although faint, can be seen just to the north of the Lagoon Nebula.

6 The Orion Nebula is thought to be a birthplace for stars.

– sparse clusters of stars of the T Tauri type. These are to be discussed in further detail later, but the T Tauri stars are very young stars which have only just begun their life.

The globular clusters are a class of objects on their own. They may contain tens of thousands of stars in a sphere of space just a few light years in diameter. The globular clusters form a halo around the nucleus of the galaxy away from the plane and some of these are quite bright objects. Examples can be seen in the constellations of Hercules M13 and in Tucanae NGC 362. Because the stars in a globular cluster are so compact, a small telescope will only reveal the stars around the outside. Quite a large telescope is required to resolve the central regions. Through a binocular a globular cluster will appear rather like a ball of cotton wool. The globular clusters seem to be highly evolved, there are no signs of gas clouds or dust. the stars which comprise these clusters are mostly very old.

Spectroscopic studies of stars in various parts of the galaxy have come up with some interesting results about their composition and their possible formation. The studies have shown that there are two basic types of star in our galaxy, referred to as population I and population II. Population I type stars are found in the plane of the galaxy and comprise by far the greatest mass of stellar material in the galaxy (there are large amounts of gas not taken into account here). Population II type stars are found away from the plane of the galaxy, chiefly in globular clusters.The difference between the two types lies mainly in their chemical make-up. All stars have hydrogen as a basic component and this may be as high as 99·5%. It is not so much the hydrogen that we are interested in but the other components, Helium and heavier elements. Population I stars contain far more heavier elements than do the population II stars and are also much younger. The reason for these basic differences is quite simple. Population II stars were formed away from the plane of the galaxy where there is very little material. The stars were formed from the most abundant element at that time, hydrogen. The stars formed under these conditions would be small in mass terms and their subsequent evolution very slow. Whilst this evolutionary path was being followed by the small, slow population II type stars, the population I stars were forming in a region of high density of gaseous material giving rise to very massive stars traversing the evolutionary path very rapidly. Some of these massive stars burn out in just a few million years, some explode and eject a proportion of the heavy elements back into the surrounding space to be reformed into other stars. In this way the amount of heavier elements in the surrounding space is increased. The stars that we observe in the plane of the galaxy are much younger and have a higher proportion of heavy elements than the stars which have evolved away from the plane of the galaxy, in the region of the globular clusters.

So far we have only looked at the standard types of galaxy and their components. Now we must include some of the peculiarities that are found amongst the collection of observed objects. With long exposure photography and the high resolution obtained with large modern telescopes comes the observation of a multitude of unusual types. The first of these are irregular structures which seem to have been affected by the gravitational influence of nearby galaxies. Next comes the so called 'cartwheel' galaxies which have a bright outer halo, giving the impression of a cartwheel – the halo forming the rim of the wheel and the spiral arms if present forming the spokes. So many of these first two have been found that there is now a special catalogue dealing solely with them.

The third type of galaxy is named after the astronomer Carl Seyfert who first noted the unusual features in 1943. The Seyfert galaxies appear to be quite normal at a first glance, but further observation reveals their nuclei to be far from normal. The nucleus of a Seyfert type of galaxy looks like a star. What causes this is not yet known but whatever it is the power emitted by the nucleus alone is far greater than the rest of the galaxy put together. Present theory suggests that the central regions of a Seyfert are undergoing a violent explosion causing the emission of not only large amounts of light but also radio waves. Many of the Seyferts are included in the list of radio galaxies. I have observed an 11th magnitude Seyfert galaxy and my description would be 'fairly bright object with uniform surface brightness, but extremely bright stellar nucleus'.

The next type of object is the radio galaxy. These are normal spiral, elliptical or irregular in shape but have an enormous radio output. Reasons for the output at radio wavelengths are unclear to say the least but they may involve violent explosive events in the nucleus or the emission from the collisions of high energy particles. In some cases the radio emission seems to arise from a source some distance from the galaxy itself. These are known as the radio lobes, often associated with an optical counterpart rather like a jet of material ejected from the nucleus of the galaxy.

Finally we come to the quasars – quasi stellar objects or Q.S.O.s for short. It was the radio astronomers that found these objects during their search for radio sources in space, but the surprise came when optical astronomers tried to link the radio sources with an optical object, i.e. an object that can be observed visually with a telescope. All that was visible was a small blue star-like object, very faint and seemingly a point source of light. Radio astronomers, checking the amount of radiation being emitted from a single Q.S.O. found that the output was as high as from a normal galaxy. Over a longer period of observation it was found that the Q.S.O.s varied their output, changing the quantity of emission over a fairly short period. Here the story gets very interesting because the period of variation of any object sets a limit to its size. The optical astronomers said that a Q.S.O. is a point source of light and the radio astronomers said that a Q.S.O. emits as much energy as a normal galaxy. The variability showed that the minimum size for a Q.S.O. would be about the size of our solar system. Imagine for a minute an object the size of the solar system, with a diameter of 2 thousandths of a light year, emitting as much light as our galaxy with a diameter of 100,000 light years and containing some 100 thousand million stars. Is a Q.S.O., then, an object which is small and quite close (relatively speaking) or large and at a truly immense distance. It was here that the spectroscope made the story even more intriguing. The spectrum of a Q.S.O. showed that the lines of emission in this case had been shifted towards the red. This is called the red-shift and is an indicator of velocity in the direction away from the observer. Calculation of the velocity of many Q.S.O.s showed that the recession speeds are truly phenomenal – up to about 80% the speed of light – indicating distances not yet encountered in our normal view of the universe.

An astronomer called Edwin Hubble observed a great number of galaxies and accurately measured their red-shifts and found that the more distant objects had higher red-shifts than close objects. From his studies he found that there is a distinct correlation between red-shift and distance. Applying this to the observed measurements taken from a Q.S.O., we get the result that the Q.S.O.s are at what could be called the outer limit. It could be that the red-shift is playing tricks, or the Q.S.O.s could be playing tricks, but until vastly more data has been accrued on this subject it is impossible to determine one way or the other as to what the quasars really are. It is possible that we are observing a violent end to a galaxy in a quasar – a stage beyond the activity in the Seyfert galaxies.

Fig. 32. A quaint group of very faint galaxies, Abell 1060.

7
The Stars

Star nomenclature began with the earliest civilisations. They wrote into their history stories of their heroes, and to give immortality to the legends they assigned groups of stars to represent the figures in the legends. These developed into the present day constellations, the names of which represent the mythical figures and stars within the constellations represent certain aspects about the figures.

With modern science came the need for an easy reference system, enabling any star to be referred to quickly and simply. This came in the 1600s when Bayer adopted the Greek lettering system. The brightest star in a constellation became designated alpha, the second brightest beta and so on. The Greek alphabet is shown in Table 9.

α Alpha	ι Iota	ρ Rho
β Beta	κ Kappa	σ Sigma
γ Gamma	λ Lambda	τ Tau
δ Delta	μ Mu	υ Upsilon
ε Epsilon	ν Nu	φ Phi
ζ Zeta	ξ Xi	χ Chi
η Eta	ο Omicron	ψ Psi
θ Theta	π Pi	ω Omega

Table 9. The Greek alphabet.

This worked well for a time, but with the advent of large telescopes which could see stars far fainter than those that Bayer had encountered, specialist star catalogues were developed with the sole task of listing one particular type of star.

A second problem, to be overcome later, was that the ancient people defined no distinct boundaries to their groups of stars. This was put to rights with the reformation of the constellations by the International Astronomical Union. Each boundary was made to lie parallel to the celestial co-ordinates of Right Ascension and Declination.

Today the stars are still referred to by their Greek letter designations, but there are several more star catalogues around which are specifically used for a given type of star. Examples of these are the 'Catalogue of Variable Stars' by Kukarkin and Parenago, the 'Struve Catalogue' and the 'Aitken Catalogue of Double Stars' designated ADC.

Classification of variable stars began with a simple lettering system which started at the letter R and followed through to Z. As these letters were used up in one constellation it became necessary to extend the scale, this was accomplished by using the letters RR to RZ, TT to TZ and so on. As the size of astronomical instruments increased, this scale was soon to become too small. Again the system was revised and expanded by continuing the scale with the letters AA to AZ, BB to BZ, LL to LZ, the letter J being omitted. This gave a total combination of 334 variable stars in any one constellation. After this a new variable star is listed simply as V 335, V 336, etc.

Variable stars are stars which, because of some physical process, vary their light output. This difference between a star's maximum and minimum brightness is called the amplitude and the length of time over which the change in brightness occurs is called the period. The latter is usually measured in days.

There are two basic types of variable stars; intrinsic – those stars which vary their luminosity by some internal physical process – and extrinsic – stars whose light varies because of an external process such as the presence of a second star in the system causing the variation in light by periodically eclipsing the first as the two stars rotate around their common centre of gravity.

Intrinsic Variables

CEPHEIDS

Named after the prototype star delta Cephei, these stars are very useful because it was discovered that there is a distinct relationship between the amplitude of the light variation and the absolute magnitude or true luminosity of this type of star. A knowledge of this relationship means that if the amplitude is known, and this is observable and easily obtainable, then we can find the absolute magnitude. Since absolute magnitude is a function of distance and apparent magnitude, the distance, which is the only unknown quantity, can be calculated. We have here a method by which we can calculate distances at least across close space, to the globular clusters and nearby galaxies. The light curve of a typical Cepheid variable star shows a steep rise to maximum followed by a steady relax period to minimum, see Fig. 33.

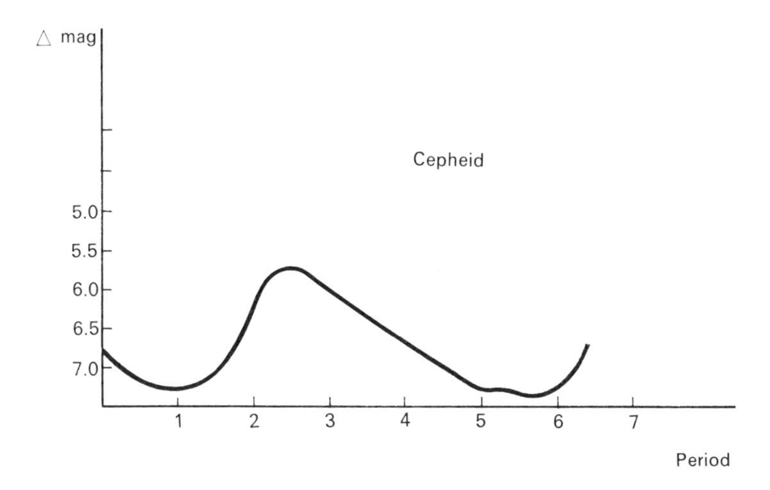

Fig. 33. Typical Cephid type light curve.

RR LYRAE

These stars are very similar to Cepheid stars, but the light curve does not rise nearly as rapidly and the drop to minimum occurs at almost the same rate as the increase. Unlike the Cepheids, RR Lyrae type stars show a stable minimum period before rising to maximum, see Fig. 34.

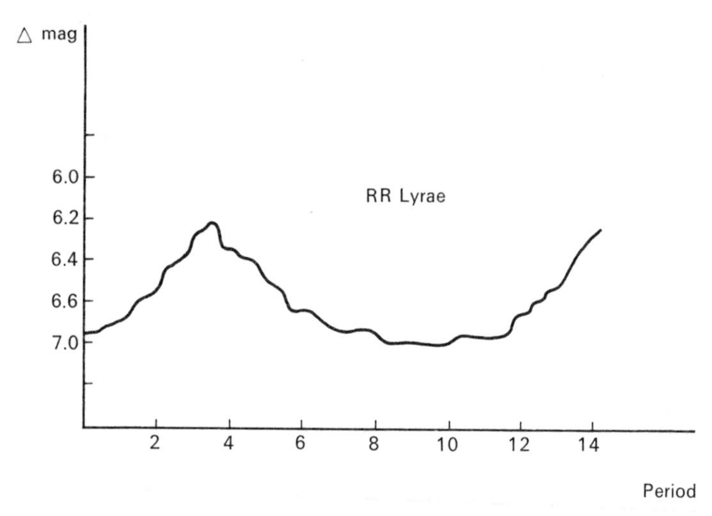

Fig. 34. RR Lyrae type light curve.

R CORONAE BOREALIS

The R. CrB stars are in one of the more peculiar classes of variable star. These stars remain at or near to maximum brightness for considerable periods of time and then suddenly drop down to minimum brightness which may be 1/1600th of the original intensity, but this is accompanied by an almost immediate rise back to maximum, this shows up well in Fig. 35.

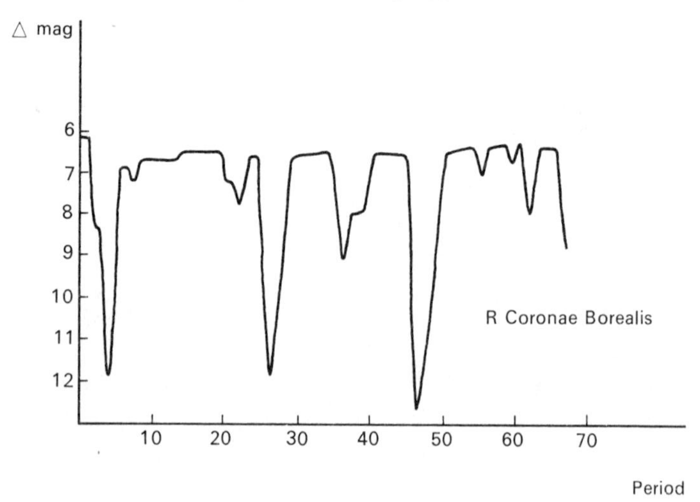

Fig. 35. R Coronae Borealis light curve.

LONG PERIOD

These stars are invariably of the red giant type of star. The period is usually fairly even and the amplitude of the variation consistent, see Fig. 36. These stars are usually called Mira type variables after the prototype star Mira Ceti.

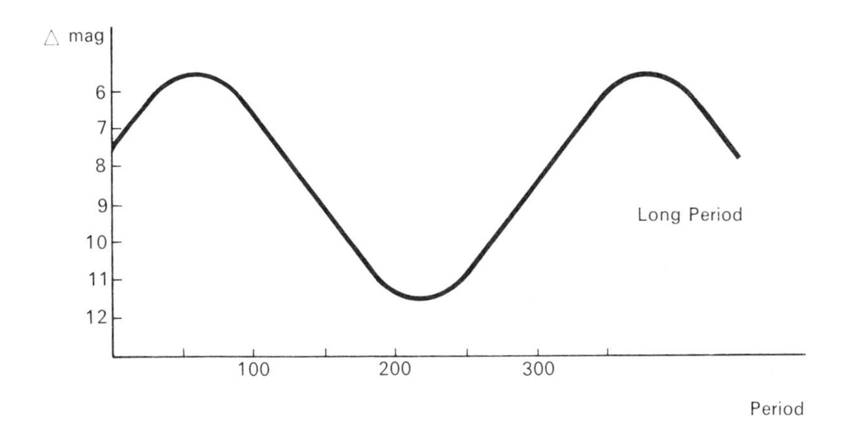

Fig. 36. Long Period or Mira type light curve.

SEMI REGULAR

These stars may fall close to any of the above classes of variable star but either their period or amplitude may vary. The variation changes only slightly between each period but over longer periods this will show that the star is in a separate class. RV Tauri is a good example of this. Whilst its period and amplitude may remain almost in phase for some periods of fluctuation, at others this star becomes completely irregular showing no regular pattern of correlation of either period or amplitude, see Fig. 37.

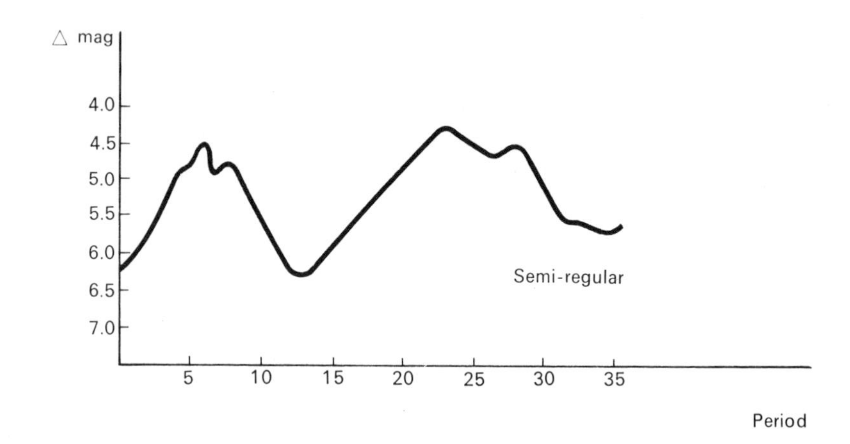

Fig. 37. Semi-regular light curve.

U GEMINORUM

These explosive variables are often called dwarf novae because of their characteristic rapid rise to maximum. The rise to maximum occurring in only a few hours and the amplitude of the variation changing by as much as four magnitudes. A typical light curve is shown in Fig. 38.

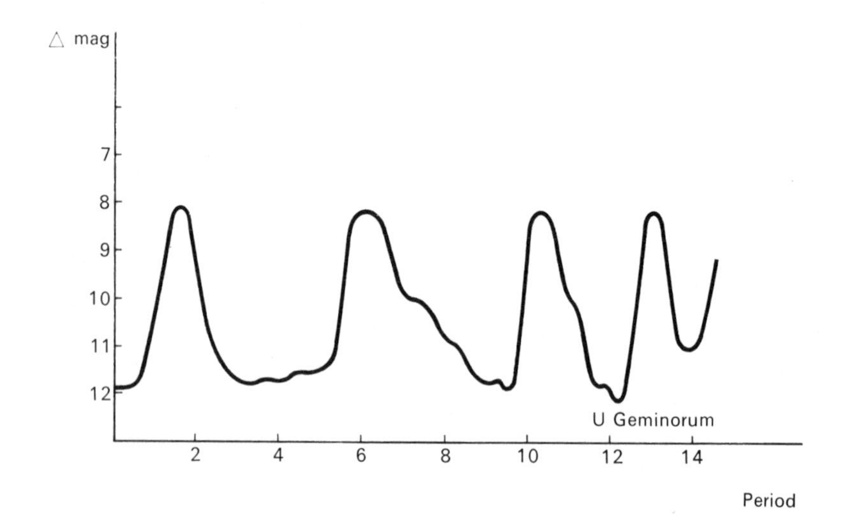

Fig. 38. U Geminorum or SS Cygni type explosive variable light curve.

IRREGULAR

Many stars fall into this class and as the name suggests there are no regular features in either the period or amplitude whatsoever.

FLARE STARS

These are again explosive variables, but with much less regularity than U Gem. types. The flares are of small amplitude and both the rise and fall are rapid, the whole cycle taking only a few hours. This makes their observation quite difficult to carry out. The work is usually done with time exposure photography to reveal the short period changes.

T CORONAE BOREALIS

Repeating novae are very explosive events and are very rare. The amplitude of these stars' variations may be as much as eight magnitudes. The reason for these variations is as yet rather uncertain, but it is possibly due to the accumulation of hydrogen at the star's surface. The extra hydrogen is drawn away from a close companion star, and as the concentration of hydrogen increases, it begins to burn near the star's surface causing a rapid increase in luminosity.

NOVAE

True novae are violently explosive eruptions within the depths of a star causing an increase in the radius of the star and a subsequent increase in luminosity. The output of energy from a nova is thought to be in the range of a million million times the Sun's normal output of energy, during its brief rise to maximum. These events are rarely fatal to the star but a great deal of the star's mass is lost during an event like this.

SUPERNOVAE

A star will not survive the effects of this, the most eruptive event known in the life of a star. The causes are most probably the result of a star being overweight. A supernova will literally tear a star to pieces. The rise in luminosity occurs over a very short period, usually in a few hours and the increase in output may be as much as twenty magnitudes. Very few supernovae occur, about one per year of the most violent type, but we can observe these in external galaxies and this is where observations have been concentrated over the last few years. When one of these events occurs in an external galaxy, it usually out-shines the rest of it.

A supernova was observed by the Chinese in the year 1064 in the constellation of Taurus, The Bull, when it was visible for several weeks with the naked eye. This has now formed what is known as M1 – number one in the Messier catalogue of diffuse objects. All that is left of the star is a huge cloud of expanding gas and dust and a very small compact star at the centre known as a neutron star. The star is also a 'pulsar', sending out pulses of light and radio emission rather like an accurate clock.

Eclipsing stars

This next group of variable stars are all part of the group of extrinsic variables. The variation in their light output is caused by an external process. All known cases are binary systems where there are two stars revolving around a common centre of gravity. At some time during the travel in their respective orbits one of the stars is eclipsed by the other causing the variation in light output. The period of the variation is a direct consequence of the orbital period of the stars.

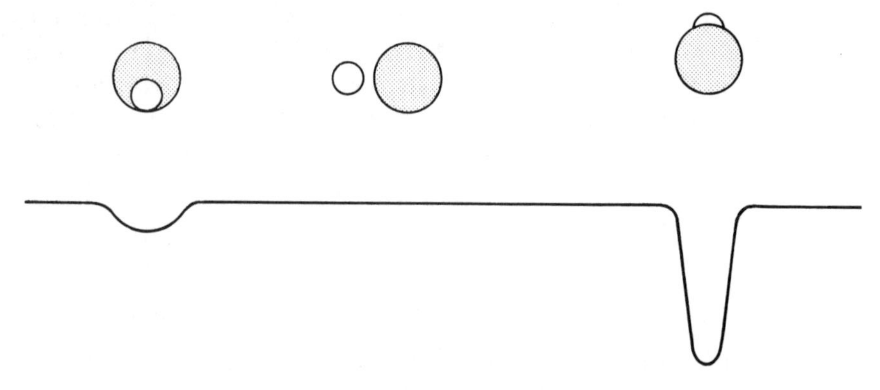

Fig. 39. Schematic representation of a binary variable star or extrinsic variable. The variation in light occurs as the stars eclipse each other.

ALGOL TYPE

These are one of the more common types of extrinsic variable star. The light curve exhibits two deep minima with a shallow minimum between, see Fig. 40.

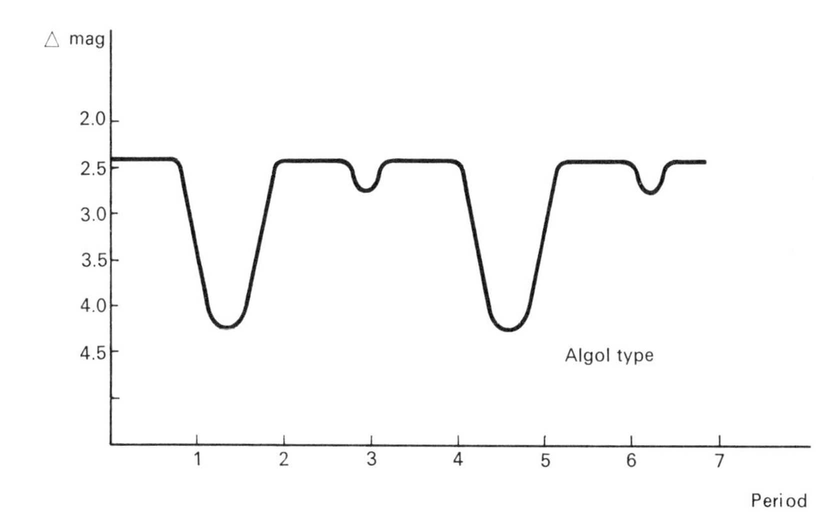

Fig. 40. The light curve of a typical Algol type variable star.

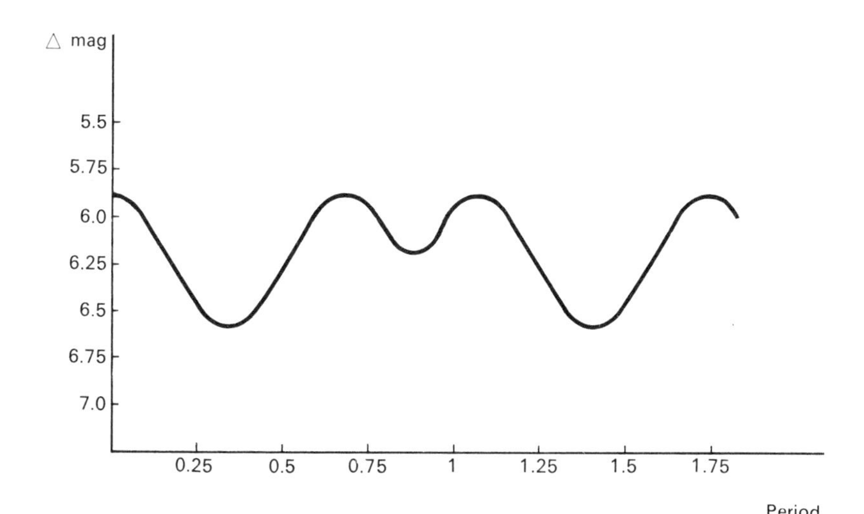

Fig. 41. The light from a Beta Lyrae type star constantly varies over the orbital period of the two stars.

BETA LYRAE TYPE

In this class the two stars are so close together that the stars are drawn into ellipsoidal shapes by the gravitation of the other star. During the cycle or orbital period the light varies continuously because the stars are changing their shapes as seen edge on to the system, see Fig. 41.

Double Stars

Double star nomenclature is based on a much simpler system. The star is referred to by its Greek letter or a specialist catalogue number such as the Aitken number. There are several basic types of double star, the first type is the optical doubles. Optical doubles are double only because two stars which are lying at different distances from the Earth are in the same line of sight, the stars thus appear to be close together but there is no gravitational attraction between them. Stars which are physically close together and are connected by mutual gravitational force are usually referred to as binary stars. These fall into two classes; those which are optical binaries and can be observed optically, and spectroscopic binaries that lie so close together that they can only be detected as binaries through the spectroscopic Doppler effect which we shall come on to later. The importance of observing double stars lies in the fact that, if the two are physically connected and we can deduce the true distance of the pair, knowing the period at which the two stars take to complete one orbit and the separation of the two stars it then becomes possible to determine approximately the masses of the stars. Spectroscopic binary stars were discovered from studies of what were thought to be individual stars. The spectroscope showed the normal dark absorption lines in the spectrum, but the peculiarity was that all the lines were doubled. This is known as the Doppler shift which is due to the velocity of the light sources towards or away from the observer. As the stars in a binary system rotate around their common centre of gravity at certain times one of the stars will be approaching the observer whilst half a period later that star will be receding from the observer. This shows up as a characteristic shift of the lines towards the red or blue end of the spectrum, red for receding and blue for approaching objects, see Fig. 42.

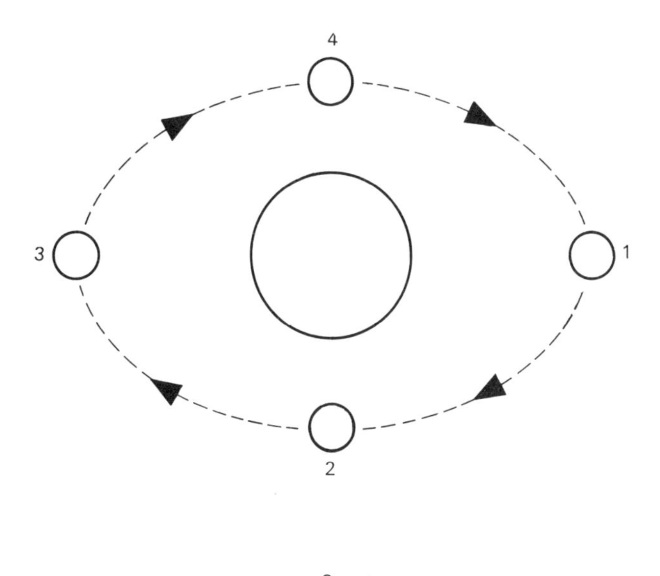

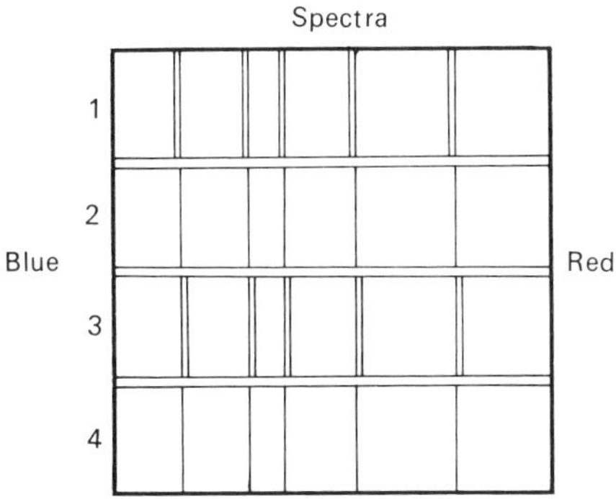

Fig. 42. The change occurring in the spectrum of a very close binary system, too close to be detected with a visual telescope. At points 1 and 3, the spectral lines have been doubled, the extra line moving to either the red or blue end of the spectrum corresponding to approach or recession of the second star.

Stellar Evolution

In general terms all stars are hot gaseous bodies. The stars shine because of this intense energy given off at the surface. Amongst the radiation emitted by a star there is a small amount of radio energy, a considerable amount of infra-red or heat, visual light and ultra-violet radiation. The spectroscope, discussed in some detail earlier, is a tool that can be used to study the light emitted by an object. The spectroscope will reveal details about the surface temperature and pressure of a star, the composition, and in some cases the atomic abundances present at the surface. A study of stellar spectra shows that all spectra can be incorporated into a fairly simple scale of characteristics. The scale can be easily remembered by the mnemonic, 'Oh Be A Fine Girl and Kiss Me Right Now or Soon'. The scale is in order of decreasing temperature and the characteristics are shown in Table 10.

Temperature (°C)	Spectroscopic Type	Characteristics
80,000+	O	Ionised and neutral helium and ionised metals.
20,000	B	Neutral helium, ionised metals and some hydrogen.
10,000	A	Mainly hydrogen with some ionised metals.
7,500	F	Hydrogen weaker with neutral metals.
6,000	G	Neutral metals and weak hydrogen with prominent calcium lines.
4,500	K	Neutral metals.
3,000	M	Neutral metals and molecular lines.

Table 10. Scale of star characteristics as observed by spectroscope.

The groups R, N and S are sub-groups of the spectral types K and M, referring to discrete differences in the metallic elements found in these stars. Groups R and N show carbon and carbon-nitrogen molecular bands and group S shows zirconium oxide and titanium oxide bands. There are three classes of objects which do not come into the previous scale – types W, P and Q. W stands for the Wolf-Rayet stars – extremely hot stars with surface temperatures of about 100,000 °K – type P stands for the gaseous emission nebulae and type Q is used for novae stars. The next logical step to take is to correlate the surface temperature with true luminosity. This can be done with a simple diagram which compares spectral type with absolute magnitude as in the Hertsprung-Russell diagram in Fig. 43. This shows the luminosity expressed as magnitude along the

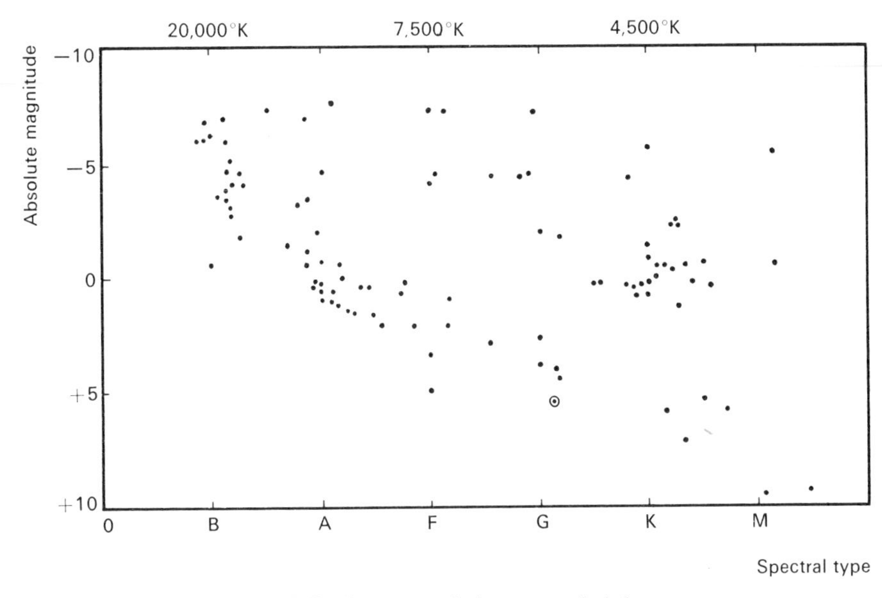

Fig. 43. H-R diagram of about 100 bright stars.

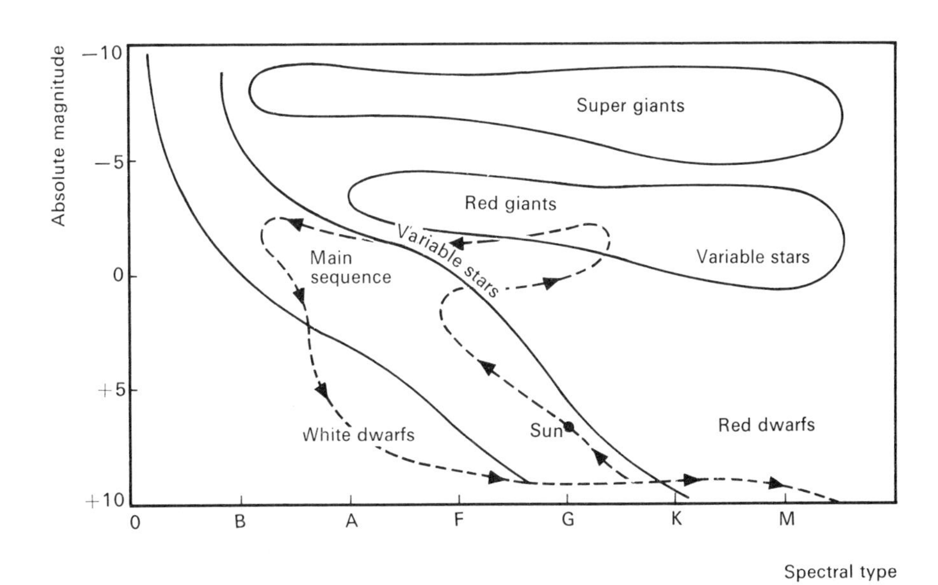

Fig. 44. H-R diagram showing areas in which stars of a certain class can be found. The dotted line shows roughly the path taken during the life of a low mass star.

vertical axis and spectral type on the horizontal axis, of the 100 brightest stars observable from the Earth.

In this form it is evident that the majority of stars lie roughly on a line extending from the bottom right to the top left of the diagram. As the number of stars included in the H-R diagram is increased to include red giant stars, variable stars and members of double star systems, and stars for which the evolutionary age is known, it becomes possible to plot a single star's evolutionary path through the H-R diagram. This, together with the areas in which stars of certain types can be found, is shown in Fig. 44.

As has already been discussed, the stars' energy is created by the fusion of atoms, forming heavier elements. We are concerned firstly with the main sequence stars, those that lie in the boundary of 'main sequence' in Fig. 44. These stars are burning hydrogen to form helium in a central core whose temperature is determined by the mass of the star. The higher the temperature of the star the larger the core will be in which the vital thermo-nuclear reactions can take place. It should also be obvious that a star cannot keep on burning hydrogen, that it will eventually run out of fuel. Extensive computer programmes have been formulated to determine to what point the concentration of hydrogen must fall before the conversion of hydrogen to helium will cease. The level which to all intents and purposes would suggest that a star is leaving the main sequence would occur as the hydrogen concentration in the core dropped to below 10%. From this it is possible to determine the time that a star of say one solar mass would remain in the main sequence bracket. For the Sun, i.e. one solar mass, the main sequence lifetime is of the order of 10^{10} years. As the mass of the theoretical model is increased, one would assume that the lifetime would also increase, but this is not so. As the mass is increased, the core temperature and the volume of the core are also increased. The main sequence lifetime of a star of two solar masses is slightly shorter than our Sun's lifetime. If we now go to a star of thirty solar masses the main sequence lifetime is only about a million years. Our Sun will last about ten thousand times longer than this.

As the concentration of hydrogen in the core is reduced, the energy output is reduced due to the fact that the reactions begin to slow down. As the star thus cools slightly, the core and the surrounding layers of the star start to contract, bringing about a gradual rise in the core temperature. The contraction and increase in temperature continue in a stable manner until sufficient density and temperature are reached, when a subsequent reaction can begin. This is the fusion of helium into berilium and lithium. This reaction begins at the core of the star and as the temperature increases again, from the induced nuclear activity, hydrogen may start to burn in a shell around the helium-burning core. This has the effect of pushing the outlying layers of the star away from the core and a gradual expansion of the outer layers ensues. With the increased radius and hence surface area, less energy has to escape from a given area of the surface of the star and whilst the core temperature may reach several hundred million degrees Kelvin, the surface temperature may be as low as 4000°, appearing to be very large and, compared with other stars, quite cool. An excellent observational example of this type of star can be found in the constellation of Orion. Betelgeuse or alpha Orionis has a diameter of 250 million miles and a surface temperature of 3000°. From the appearance of such stars, they have earned themselves the common name 'red giants'.

The next stage in the life of a star is the death stage. Depending on the mass of the star the duration of this stage can last for any length of time. A low mass star, on reaching the stage where the concentration of heavy elements is so high that no more nuclear synthesis can take place, will start to contract. As the contraction ensues, the surface temperature rises. The star becomes very dense and once an equilibrium is reached where the star can contract no more it will begin to cool down, losing its last reserve of energy – gravitational energy. In the more massive stars this process will be slightly prolonged because with more mass a further bout of nuclear reactions may suddenly start, producing elements as heavy as iron. This phase of a star's life is usually referred to as the 'white dwarf stage' from the appearance of a star during the phase. As the star loses its energy and cools down it passes briefly through the 'red dwarf stage' and from there to the dead star state.

Stars are not always found as separate units or concentrations of mass but more usually as part of a cluster. Even if a star is not apparently directly connected to a cluster there is a good chance that it is a rogue star that has escaped from a cluster at some time in its history.

Nebulae

All star formation begins in a large cloud of gas that we call a nebula and from this nebula, which may contain many millions of solar masses, small condensations occur to form the stars that we now see. Star formation from a nebula has been observed in the Orion nebula where discrete sources of infra-red radiation have been found suggesting that very young stars are present.

There are three observable types of gaseous nebula and it is quite common to find all three types in the same nebula system. The dark absorption nebulae are so-called because they absorb light, blocking out the light of stars lying beyond them. These clouds form the 'coal sacks' of the southern milky way. The bright nebulae can be of either the emission type or the reflection type depending on the way in which the light from the nebula reaches the observer. In the emission types the light is emitted from particles within the nebula rather in the same way as the light from a neon strip light, local stars embedded in the nebula causing the emission to occur. The reflection nebulae, as the name suggests, are visible by reflected light. These act like mirrors, reflecting the light of stars lying between the nebula and the observer.

Although the reflection nebulae may at a first glance appear similar to the emission nebulae, the spectroscope will reveal that the characteristics of the spectrum of each kind will differ. The spectrum of a reflection nebula is basically the same as the star from where the light originated.

The last type of nebula that must be mentioned is the planetary nebula. Planetary nebulae are the remnants of a nova type explosion when mass is lost from a star or multiple star system causing a cloud of gas to expand around the whole system. The shell, excited by radiation from the central system, glows in the same way as an emission nebula. The planetary nebula get their name from their likeness to the image presented by a planet, i.e. a dense solid object. These nebulae may take a variety of forms around the central object due to uneven ejection of material from the nova. One of the brighter examples can be found in the constellation of Lyra – the Ring Nebula. In this there has been uniform ejection for a limited period of time. The expanding shell of gas, seen from a distance, takes the appearance of a brightish ring of material around a central star which is of the 15th magnitude. The nebulosity is of about magnitude nine and can be seen with a 2 in. (5 cm.) telescope, but to see the ring shape of the nebula requires at least an 8 in. (20 cm.) telescope.

Appendix I Double stars

Name	R.A. hr ′	Dec. ° ′	Vis. Mag. A b	Sep ″	Notes
αUMa	01 48	89 02	2·0–9·0	18	optical
ζAnd	02 00	42 06	2·3–5·1	13	no change
ηCMi	07 25	07 03	5·3–11·3	4·1	difficult
ζVel	08 07	−47 12	2·2–4·8	41	
ζCen	12 38	−48 41	3·1–3·2	0·3	very difficult
θVir	13 07	−05 16	4·4–8·6	7·2	
εBoo	14 42	27 17	2·7–5·1	2·9	red and blue pair
σCrB	16 12	33 59	5·7–6·7	6·2	slow binary
εLyr[1+2]	18 42	39 37	4·7–4·5	208	double double
ε[1]			5·1–6·0	2·7	
ε[2]			5·1–5·4	2·3	
βCyg	19 28	27 52	3·2–5·4	34	yellow and purple
ηPsA	21 58	−28 42	5·8–6·8	1·6	very close
βCep	21 28	70 20	3·3–8·0	10	
δCep	22 27	58 10	var–7·5	41	A is cepheid variable
βPsA	22 28	−32 36	4·4–7·9	30	
94Aqr	23 16	−13 44	5·3–7·5	13	yellow and blue

Appendix II Variable stars

Name	R.A. hr ′	Dec. ° ′	Mag. range min–max	Period days	Type
T And	00 19	26 43	7·7–14·3	280	Mira type
ζ Pho	01 06	−55 31	3·6–4·1	1·6	eclipsing
UV Cet	01 36	−18 13	6·8–12·9	——	flare star
O Cet	02 16	−03 12	1·7–10·1	331	Mira type
β Per	03 04	40 46	2·2–3·5	2·8	Algol
AE Aur	05 13	34 15	5·4–6·1	——	Irregular
T Mon	06 22	07 07	5·8–6·8	27	Cepheid
T Pyx	09 02	−32 11	7·0–13·0	——	Nova (recurrent)
SY UMa	09 52	50 03	5·1–6·0	?	Unknown
δ Lib	14 58	−08 19	4·8–6·1	2·3	Algol type
α Her	17 12	14 27	3·0–4·0	?	Semi-regular

VW Dra	17 56	54 50	7·0–8·0	?	Unknown
RR Lyr	19 23	42 41	7·0–8·0	0·6	RR Lyr type
Y Pav	21 19	−69 57	5·7–8·5	233	Semi-regular
δ Cep	22 27	58 10	3·6–4·3	5·3	Cepheid

Appendix III Star clusters

Name	R.A. hr ′	Dec. ° ′	Mag.	Notes
M31	00 40	41 00	4·8	The great Spiral, galaxy in Andromeda
NGC 869	02 17	56 55	4·5	The double cluster in the sword handle of Perseus
NGC 884	02 20	56 53	4·5	
M45	03 44	24 00	2·0	The Pleiades open cluster
M79	05 22	−24 34	7·9	Globular cluster
M38	05 25	35 48	7·4	Open cluster in Auriga
M36	05 32	34 07	6·3	Open cluster in Auriga
M37	05 49	32 32	6·2	Open cluster in Auriga
M41	06 44	−20 41	4·6	Open cluster in Canis Major
M44	08 37	20 00	4·0	The Beehive in Cancer
M13	16 39	36 33	5·7	Hercules globular cluster
M6	17 36	−32 11	5·3	Open cluster Scorpius
M8	18 00	−24 23	6·0	The Lagoon Nebula in Sagittarius
M11	18 48	−06 20	6·3	The Wild Duck open cluster
M27	19 57	22 35	7·6	The Dumbell planetary nebula

Appendix IV The constellations

Constellation	Abbreviation	Genitive ending
Andromeda	And	–dae
Antilia	Ant	–liae
Apus	Aps	–podis

Constellation	Abbreviation	Genitive ending
Aquarius	Aqr	–rii
Aquila	Aql	–lae
Ara	Ara	–rae
Aries	Ari	–ietis
Auriga	Aur	–gae
Bootes	Boo	–tis
Caelum	Cae	–aeli
Camelopardus	Cam	–di
Cancer	Cnc	–cri
Canes Venatici	CVn	–num –curum
Canis Major	CMa	–is –ris
Canis Minor	CMi	–is –ris
Capricornus	Cap	–ni
Carina	Car	–nae
Cassiopeia	Cas	–peiae
Centaurus	Cen	–ri
Cepheus	Cep	–phei
Cetus	Cet	–ti
Chameleon	Cha	–ntis
Circinus	Cir	–ni
Columba	Col	–bae
Coma Berenices	Com	–mae –cis
Corona Australis	CrA	–nae –lis
Corona Borealis	CrB	–nae –lis
Corvus	Crv	–vi
Crater	Crt	–eris
Crux	Cru	–ucis
Cygnus	Cyg	–gni
Delphinus	Del	–ni
Dorado	Dor	–dus
Draco	Dra	–onis
Equuleus	Equ	–lei
Eridanus	Eri	–ni
Fornax	For	–acis
Gemini	Gem	–norum
Grus	Gru	–ruis
Hercules	Her	–lis
Horologium	Hor	–gii
Hydra	Hya	–drae
Hydrus	Hyi	–dri
Indus	Ind	–di

Constellation	Abbreviation	Genitive ending
Lacerta	Lac	–tae
Leo	Leo	–onis
Leo Minor	LMi	–onis –ris
Lepus	Lep	–poris
Libra	Lib	–rae
Lupus	Lup	–pi
Lynx	Lyn	–ncis
Lyra	Lyr	–rae
Mensa	Men	–sae
Microscopium	Mic	–pii
Monoceros	Mon	–rotis
Musca	Mus	–cae
Norma	Nor	–mae
Octans	Oct	–ntis
Ophiuchus	Oph	–chi
Orion	Ori	–nis
Pavo	Pav	–nis
Pegasus	Peg	–si
Perseus	Per	–sei
Phoenix	Phe	–nicis
Pictor	Pic	–ris
Pisces	Psc	–cium
Pisces Austrinus	PsA	–is –ni
Puppis	Pup	–ppis
Pyxis	Pyx	–xidis
Reticulum	Ret	–li
Sagitta	Sga	–tae
Sagittarius	Sgr	–rii
Scorpius	Sco	–pii
Sculptor	Scl	–ris
Scutum	Sct	–uti
Serpens	Ser	–ntis
Sextans	Sex	–ntis
Taurus	Tau	–ri
Telescopium	Tel	–pii
Triangulum	Tri	–li
Triangulum Australe	TrA	–li –lis
Tucana	Tuc	–nae
Ursa Major	UMa	–sae –ris
Ursa Minor	UMi	–sae –ris

Constellation	Abbreviation	Genitive ending
Vela	Vel	–lorum
Virgo	Vir	–ginis
Volans	Vol	–ntis
Vulpecula	Vul	–lae

Appendix V The brightest stars

Name	R.A. hr ′	Dec. ° ′	Dist. L.Y.	Mag. apparent	Mag. absolute
CMa	06 42	−16 39	8·7	−1·5	+0·7
Car	06 22	−52 40	300	−0·7	−5·5
Cen	14 36	−60 38	4·3	−0·2	+4·6
Boo	14 13	19 26	36	−0·0	−0·3
Lyr	18 35	38 44	26	+0·03	+0·3
Ori	05 12	−08 15	850	0·08	−7·0
Aur	05 13	45 57	45	0·09	+0·1
CMi	07 36	05 21	11	0·35	+2·8
Eri	01 35	−57 29	75	0·49	−1·3
Cen	14 00	−60 08	300	0·61	−4·3
Aql	19 48	08 44	16	0·75	+2·1
Tau	04 33	16 25	65	0·78	−0·2
Cru	12 23	−62 49	270	0·8	−3·8
Ori	05 52	07 24	650	0·85	−5·5
Sco	16 26	−26 19	400	0·92	−4·5

Appendix VI The nearest stars

Name	R.A. hr ′	Dec. ° ′	Dist. L.Y.	Mag. apparent
Proxima Cen	14 26	−62 28	4·3	10·7
Cen A	14 36	−60 38	4·3	0·0
Cen B	14 36	−60 38	4·3	1·4
Barnards star	17 55	04 24	6·0	9·5

Wolf 359	10 54	07 20	8·1	13·5
Lal 21185	11 00	36 18	8·2	7·5
Sirius A	06 42	−16 39	8·7	−1·5
Sirius B	06 42	−16 39	8·7	12·5
UV Ceti A	01 36	−18 13	9·0	13·0
UV Ceti B	01 36	−18 13	9·0	13·0
Ross 154	18 46	−23 53	9·3	10·5
Ross 248	23 39	43 56	10·3	12·2
Eridani	03 30	−09 38	10·8	3·7
L789/6	22 35	−15 36	11·1	12·2
Ross 128	11 45	01 06	11·1	11·0

Appendix VII Useful addresses

Royal Astronomical Society, Burlington House, Piccadilly, London W1V ONL, U.K.

British Astronomical Association, Burlington House, Piccadilly, London W1V ONL, U.K.

Irish Astronomical Society, 1 Garville Road, Dublin 6, Eire.

British Astronomical Association, (New South Wales Branch), 33 Cotswold Road, Strathfield, N.S.W. 2135 Australia.

Royal Astronomical Society of Canada, 252 College Street, Toronto 2–B, Canada.

Royal Astronomical Society of New Zealand, P.O. Box 3181, Wellington, New Zealand.

South African Astronomical Society, c/o Royal Observatory, Cape Province, South Africa.

American Astronomical Society, 335 East 45th Street, New York, 10017 U.S.A.

Astronomical Society of the Pacific, c/o California Academy of Sciences, Golden Gate Park, San Francisco, California, 94118 U.S.A.

Acknowledgements

Plates 1, 2, 4, 5, 6, and Figs. 25, 32: Copyright Royal Observatory, Edinburgh.
Plate 3: Copyright California Institute of Technology and the Carnegie Institution of Washington. Reproduction by permission from the Hale Observatories.
Artwork: Rod Paull, Gordon Dowland, David Dowland.
Star maps: Paul Rantzau, by arrangement with Politikens Forlag Ltd.
Typing: Mrs Linda Swift.

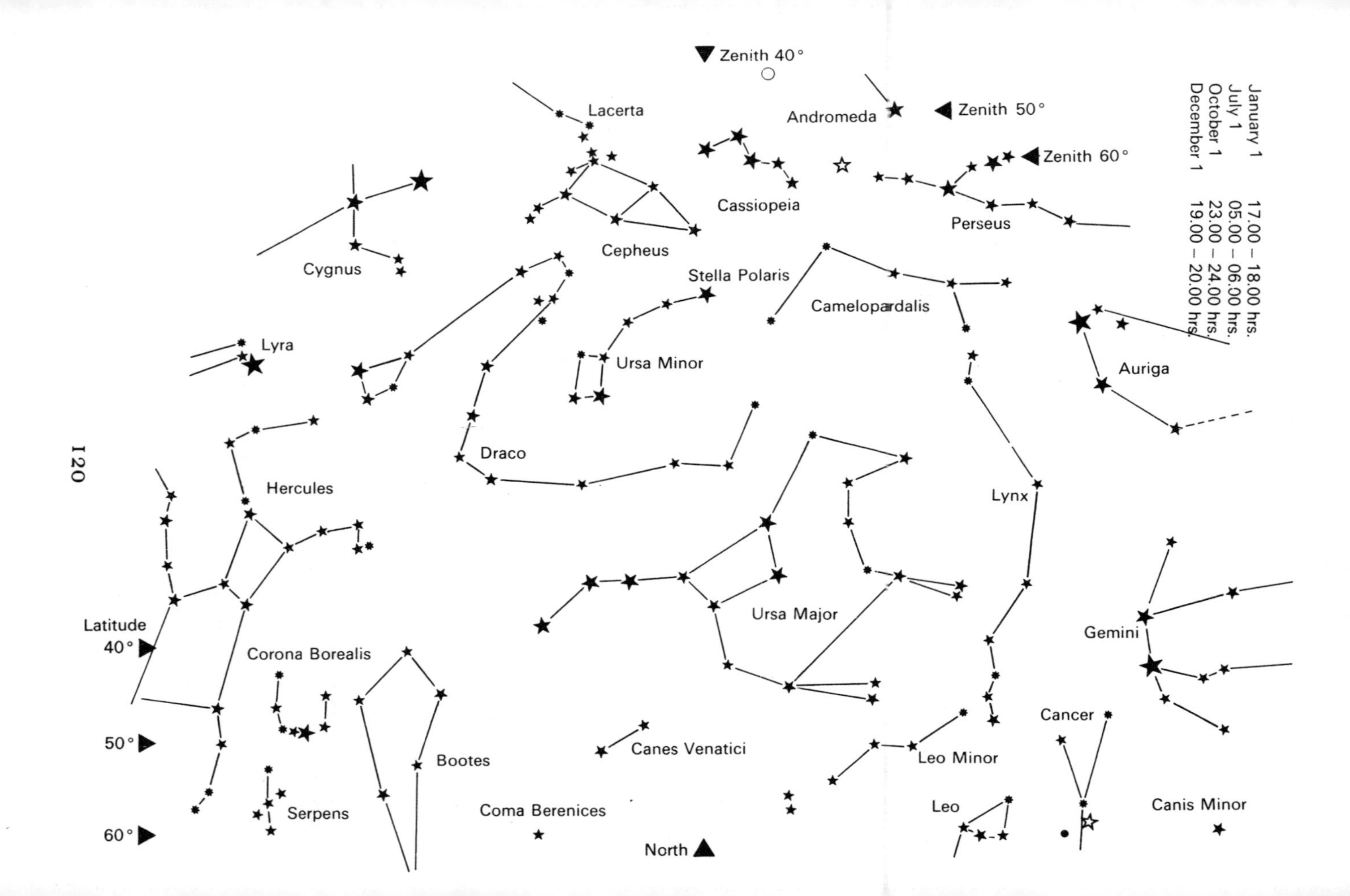
January 1 17.00 – 18.00 hrs.
July 1 05.00 – 06.00 hrs.
October 1 23.00 – 24.00 hrs.
December 1 19.00 – 20.00 hrs.
Zenith 40°
Zenith 50°
Zenith 60°
Lacerta
Andromeda
Cassiopeia
Perseus
Cygnus
Cepheus
Stella Polaris
Camelopardalis
Lyra
Ursa Minor
Auriga
Draco
Hercules
Lynx
Ursa Major
Gemini
Latitude
40°
Corona Borealis
Cancer
50°
Bootes
Canes Venatici
Leo Minor
Serpens
Coma Berenices
Leo
Canis Minor
60°
North

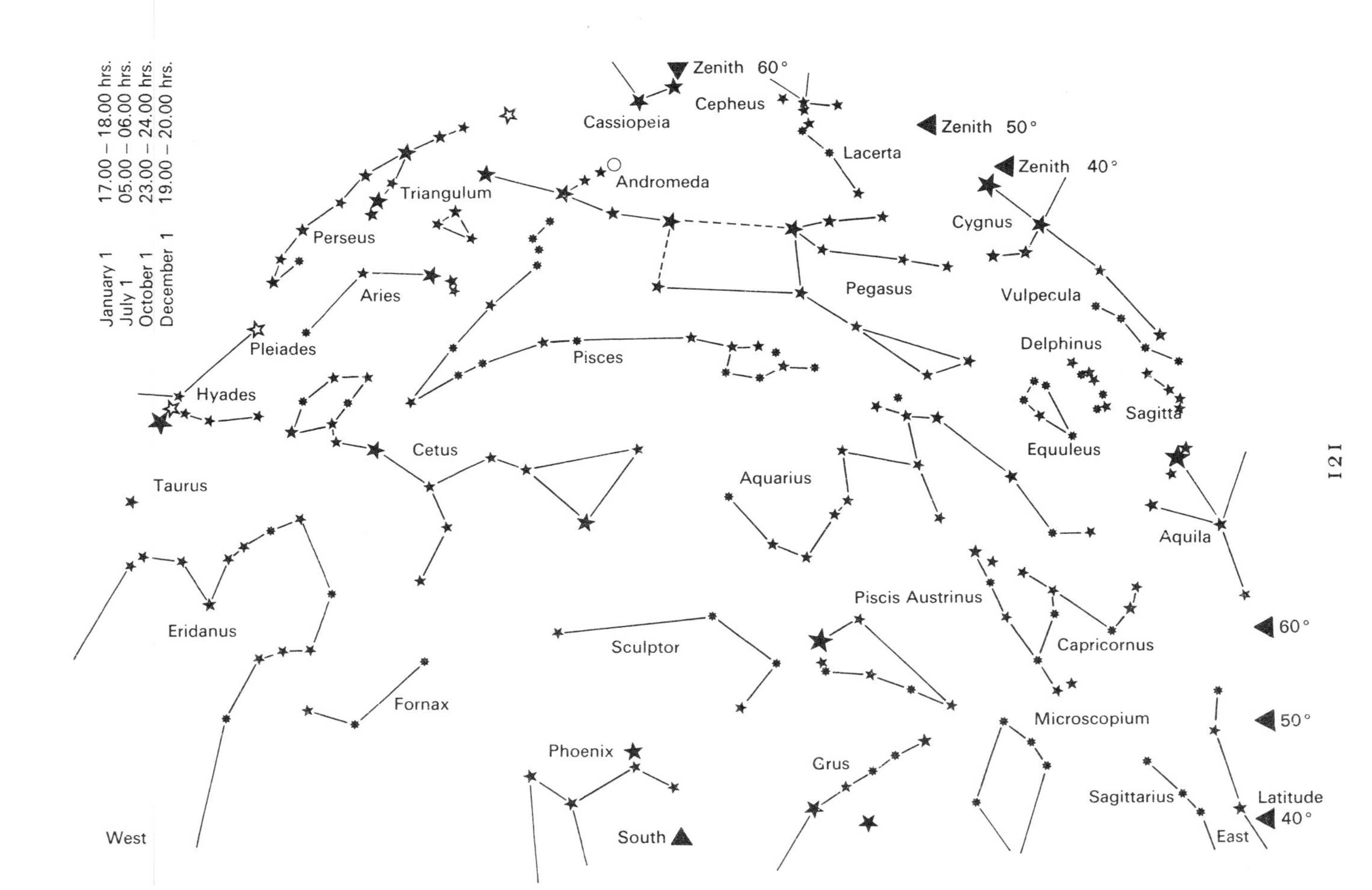
January 1 17.00 – 18.00 hrs.
July 1 05.00 – 06.00 hrs.
October 1 23.00 – 24.00 hrs.
December 1 19.00 – 20.00 hrs.
Zenith 60°
Cassiopeia
Cepheus
Zenith 50°
Lacerta
Zenith 40°
Andromeda
Triangulum
Perseus
Cygnus
Aries
Pegasus
Vulpecula
Pleiades
Pisces
Delphinus
Hyades
Sagitta
Cetus
Equuleus
Taurus
Aquarius
Aquila
Eridanus
Piscis Austrinus
60°
Sculptor
Capricornus
Fornax
Microscopium
50°
Phoenix
Grus
Sagittarius
Latitude 40°
West
South
East

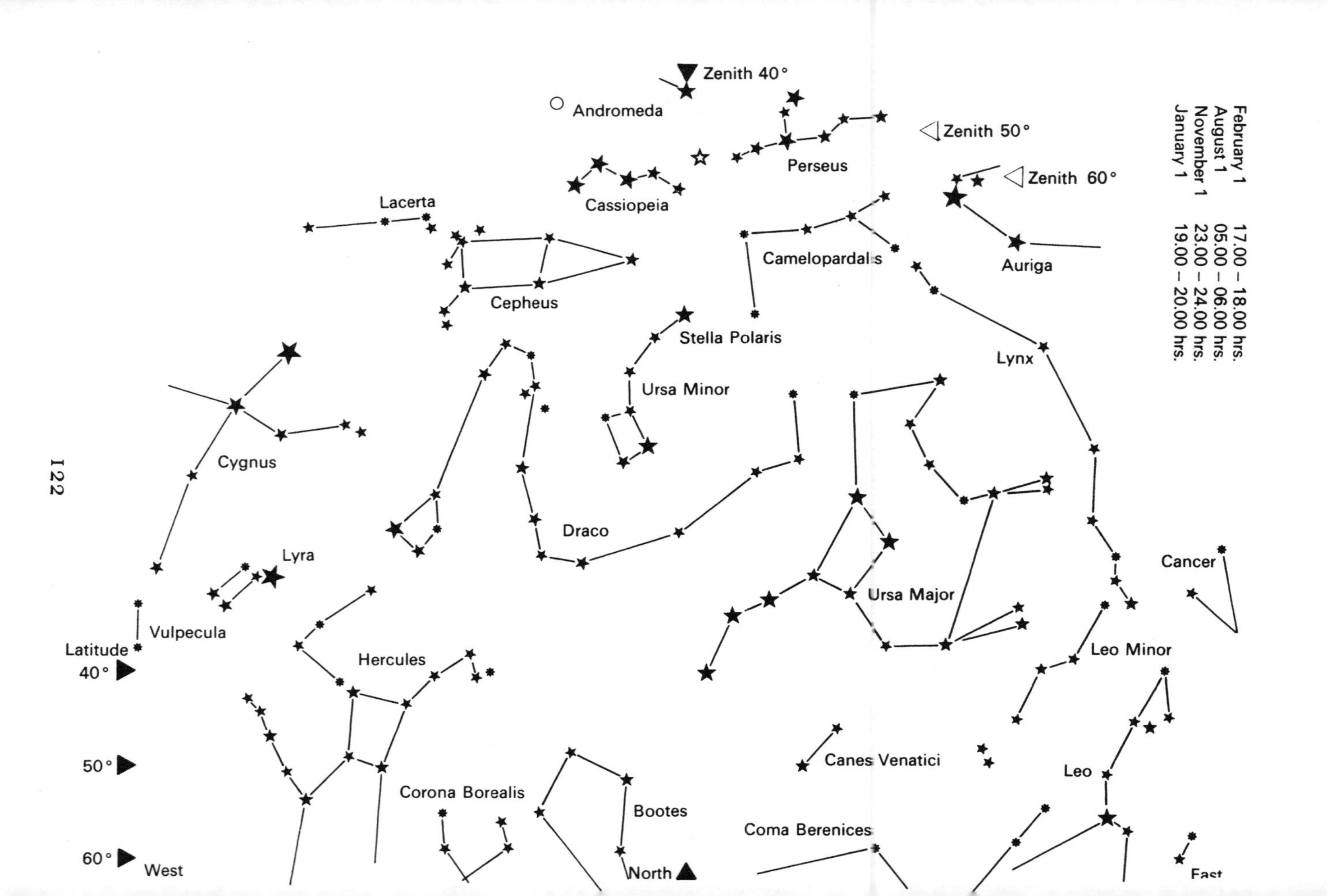
February 1 17.00 – 18.00 hrs.
August 1 05.00 – 06.00 hrs.
November 1 23.00 – 24.00 hrs.
January 1 19.00 – 20.00 hrs.
Zenith 40°
Andromeda
Perseus
Zenith 50°
Zenith 60°
Cassiopeia
Lacerta
Camelopardalis
Auriga
Cepheus
Stella Polaris
Lynx
Ursa Minor
Cygnus
Draco
Lyra
Cancer
Ursa Major
Vulpecula
Latitude
40°
Hercules
Leo Minor
Canes Venatici
Leo
50°
Corona Borealis
Bootes
Coma Berenices
60°
West
North
East

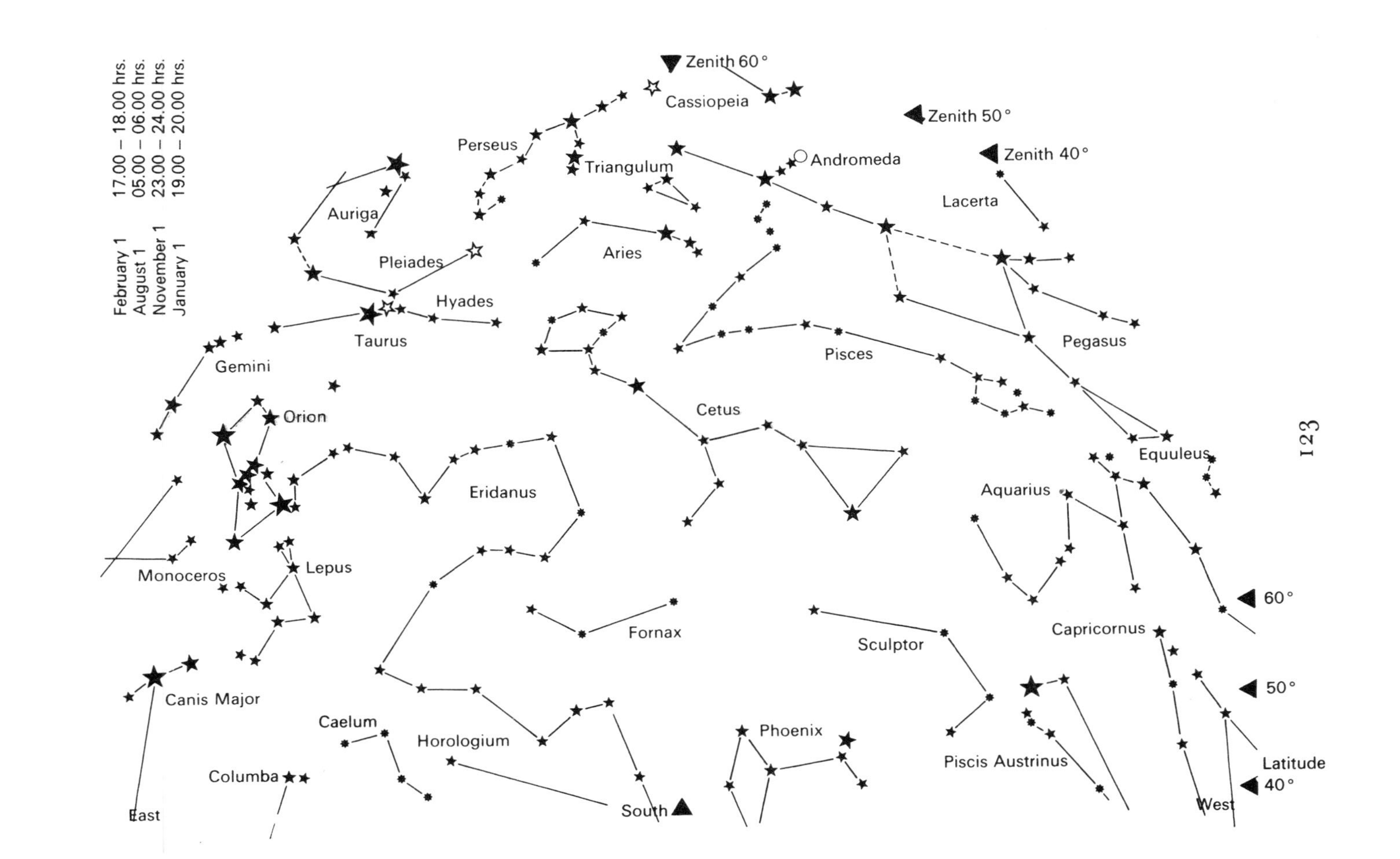
February 1 17.00 – 18.00 hrs.
August 1 05.00 – 06.00 hrs.
November 1 23.00 – 24.00 hrs.
January 1 19.00 – 20.00 hrs.
Zenith 60°
Cassiopeia
Zenith 50°
Perseus
Triangulum
Andromeda
Zenith 40°
Auriga
Lacerta
Aries
Pleiades
Hyades
Taurus
Pegasus
Pisces
Gemini
Orion
Cetus
Equuleus
Eridanus
Aquarius
Monoceros
Lepus
60°
Fornax
Sculptor
Capricornus
Canis Major
50°
Caelum
Phoenix
Horologium
Piscis Austrinus
Latitude
Columba
40°
East
South
West

Zenith 40°
Zenith 50°
Zenith 60°
March 1 17.00 – 18.00 hrs.
September 1 05.00 – 06.00 hrs.
December 1 23.00 – 24.00 hrs.
February 1 19.00 – 20.00 hrs.
Andromeda
Perseus
Auriga
Cassiopeia
Camelopardalis
Lynx
Stella Polaris
Lacerta
Cepheus
Ursa Minor
Pegasus
Draco
Ursa Major
Leo Minor
Cygnus
Latitude
40°
Lyra
Canes Venatici
Delphinus
Hercules
Coma Berenices
Leo
50°
Vulpecula
Sagitta
Corona Borealis
Bootes
60°
West
North
East

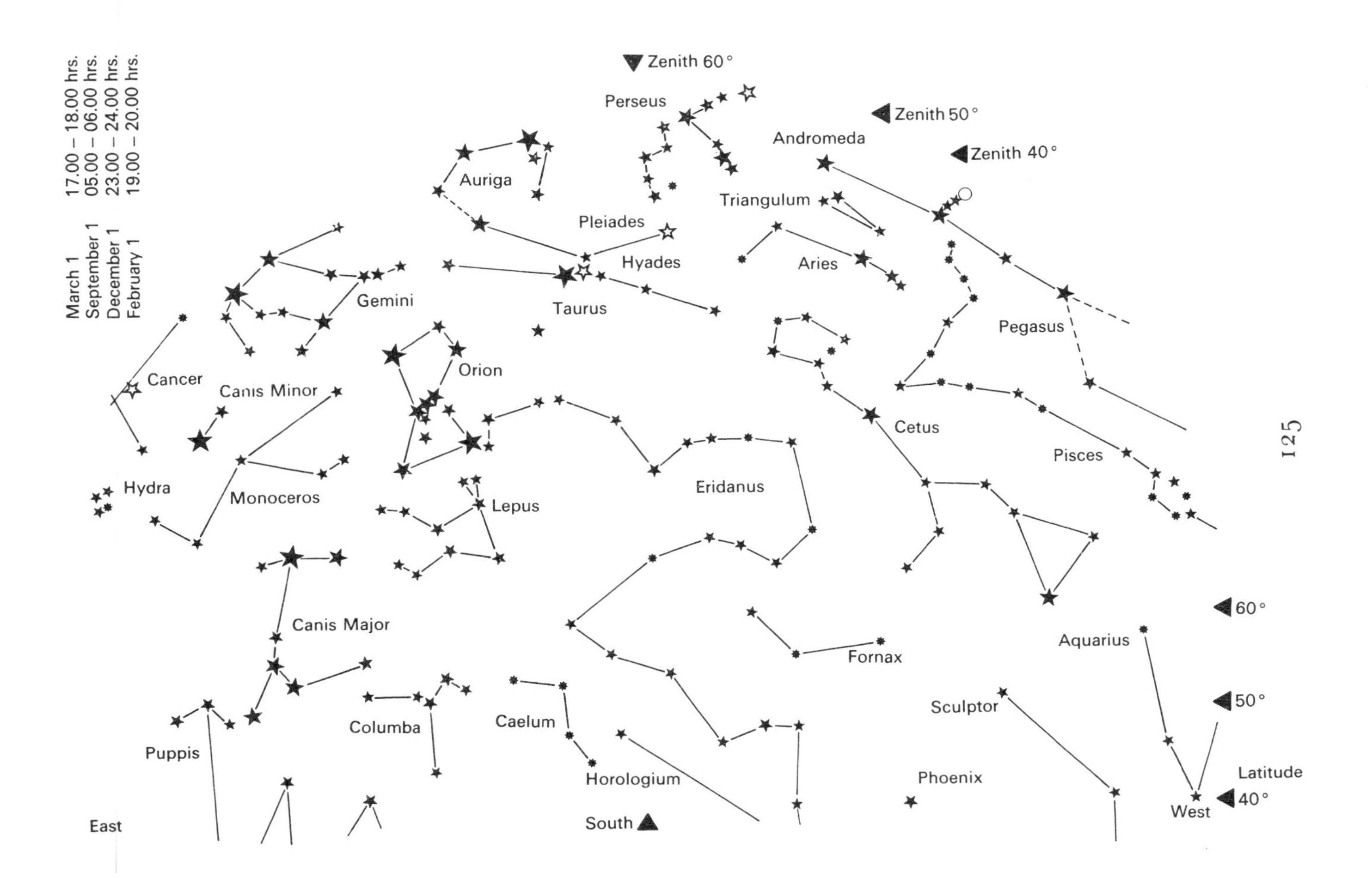

March 1 17.00 – 18.00 hrs.
September 1 05.00 – 06.00 hrs.
December 1 23.00 – 24.00 hrs.
February 1 19.00 – 20.00 hrs.
Zenith 60°
Perseus
Zenith 50°
Andromeda
Zenith 40°
Auriga
Triangulum
Pleiades
Hyades
Aries
Gemini
Taurus
Pegasus
Cancer
Orion
Canis Minor
Cetus
Pisces
Hydra
Monoceros
Eridanus
Lepus
60°
Canis Major
Aquarius
Fornax
50°
Sculptor
Columba
Caelum
Puppis
Latitude
Horologium
Phoenix
40°
West
East
South

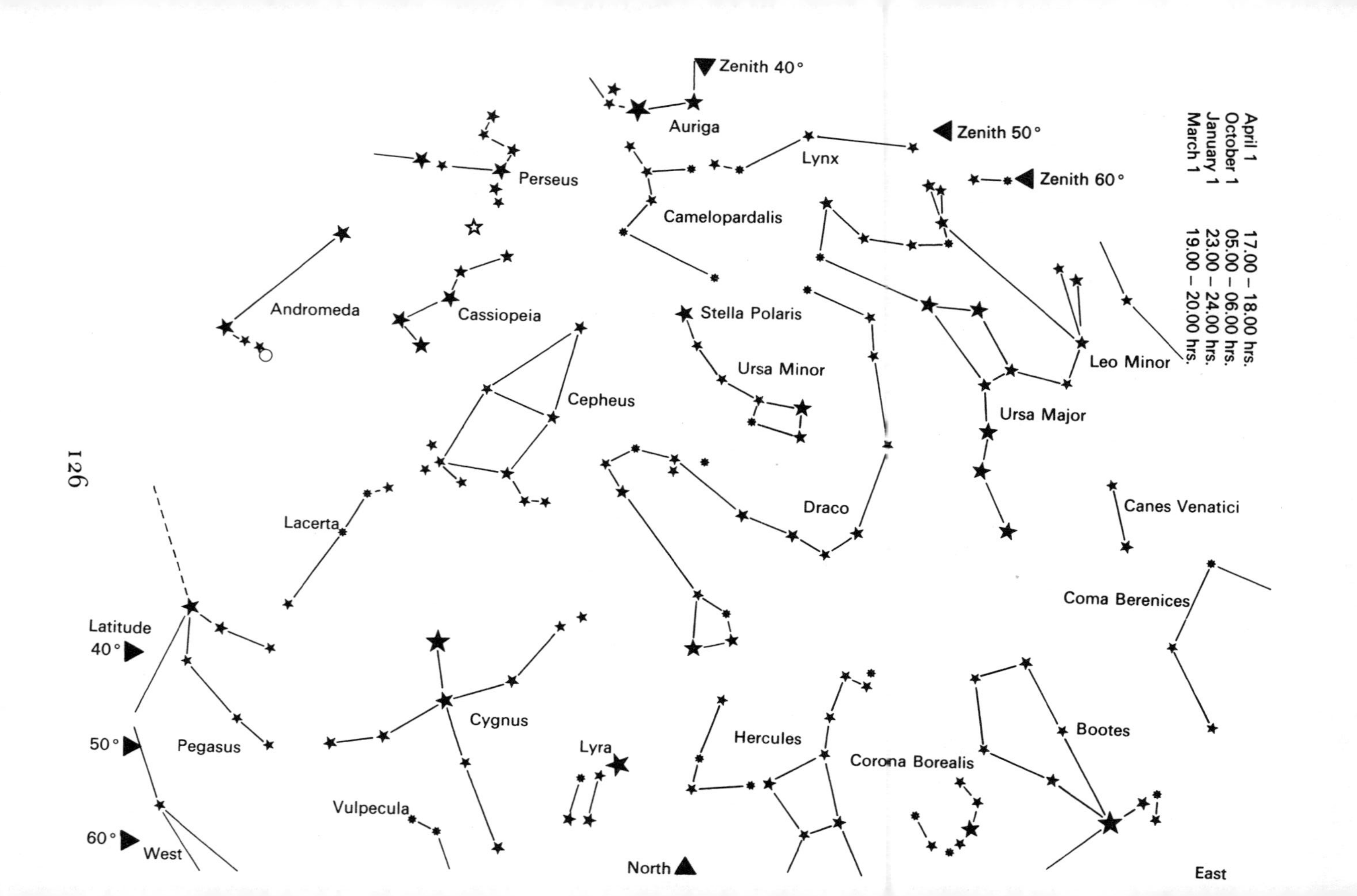

April 1 17.00 – 18.00 hrs.
October 1 05.00 – 06.00 hrs.
January 1 23.00 – 24.00 hrs.
March 1 19.00 – 20.00 hrs.
Zenith 40°
Zenith 50°
Zenith 60°
Auriga
Lynx
Perseus
Camelopardalis
Andromeda
Cassiopeia
Stella Polaris
Ursa Minor
Leo Minor
Cepheus
Ursa Major
Draco
Lacerta
Canes Venatici
Coma Berenices
Latitude
40°
50°
60°
West
Pegasus
Cygnus
Lyra
Hercules
Corona Borealis
Bootes
Vulpecula
North
East

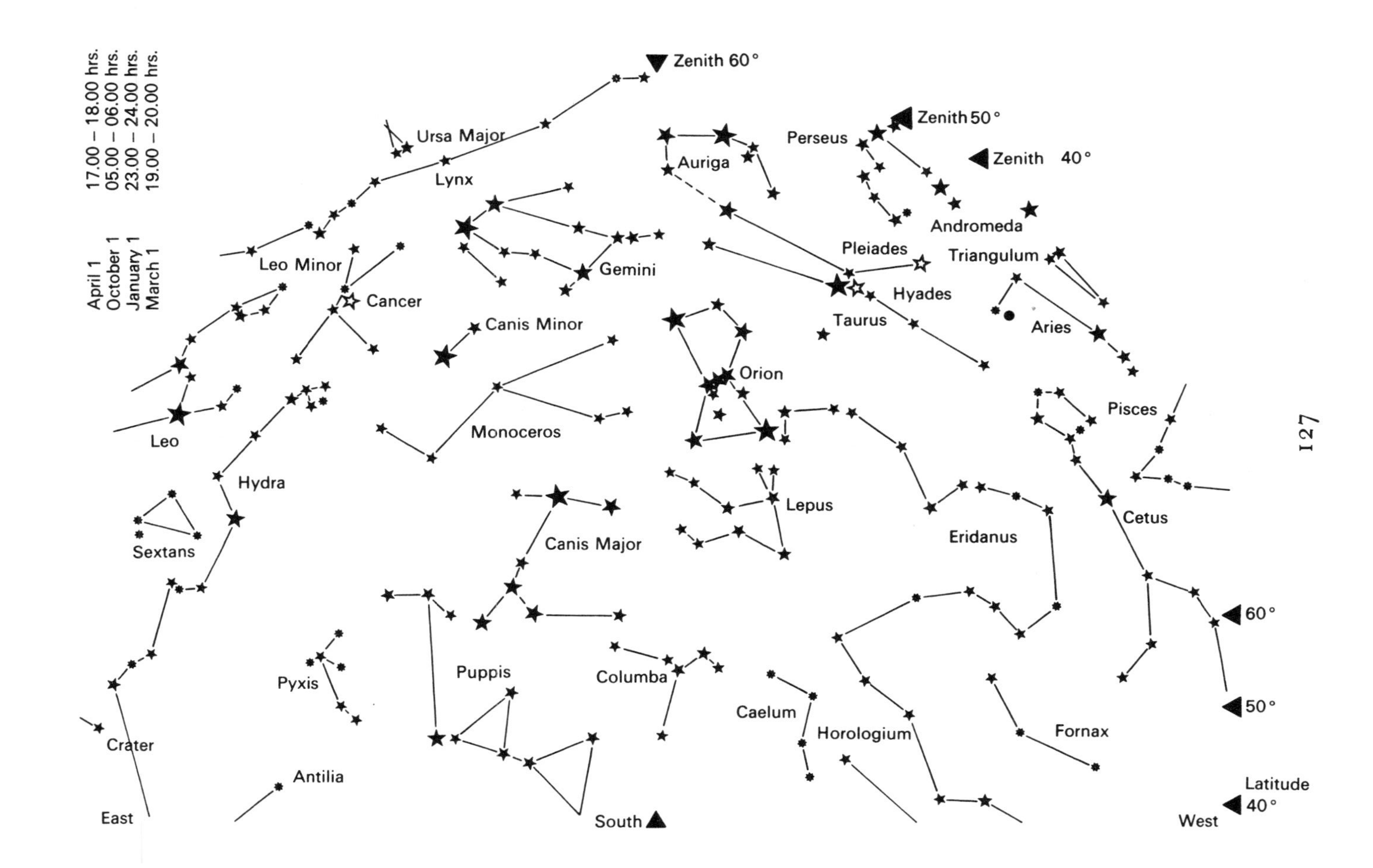

April 1 17.00 – 18.00 hrs.
October 1 05.00 – 06.00 hrs.
January 1 23.00 – 24.00 hrs.
March 1 19.00 – 20.00 hrs.
Zenith 60°
Ursa Major
Lynx
Auriga
Perseus
Zenith 50°
Zenith 40°
Andromeda
Leo Minor
Gemini
Pleiades
Triangulum
Cancer
Hyades
Canis Minor
Taurus
Aries
Orion
Leo
Monoceros
Pisces
Hydra
Lepus
Cetus
Sextans
Canis Major
Eridanus
60°
Puppis
Pyxis
Columba
Caelum
Horologium
Fornax
50°
Crater
Antilia
Latitude
40°
East
South
West

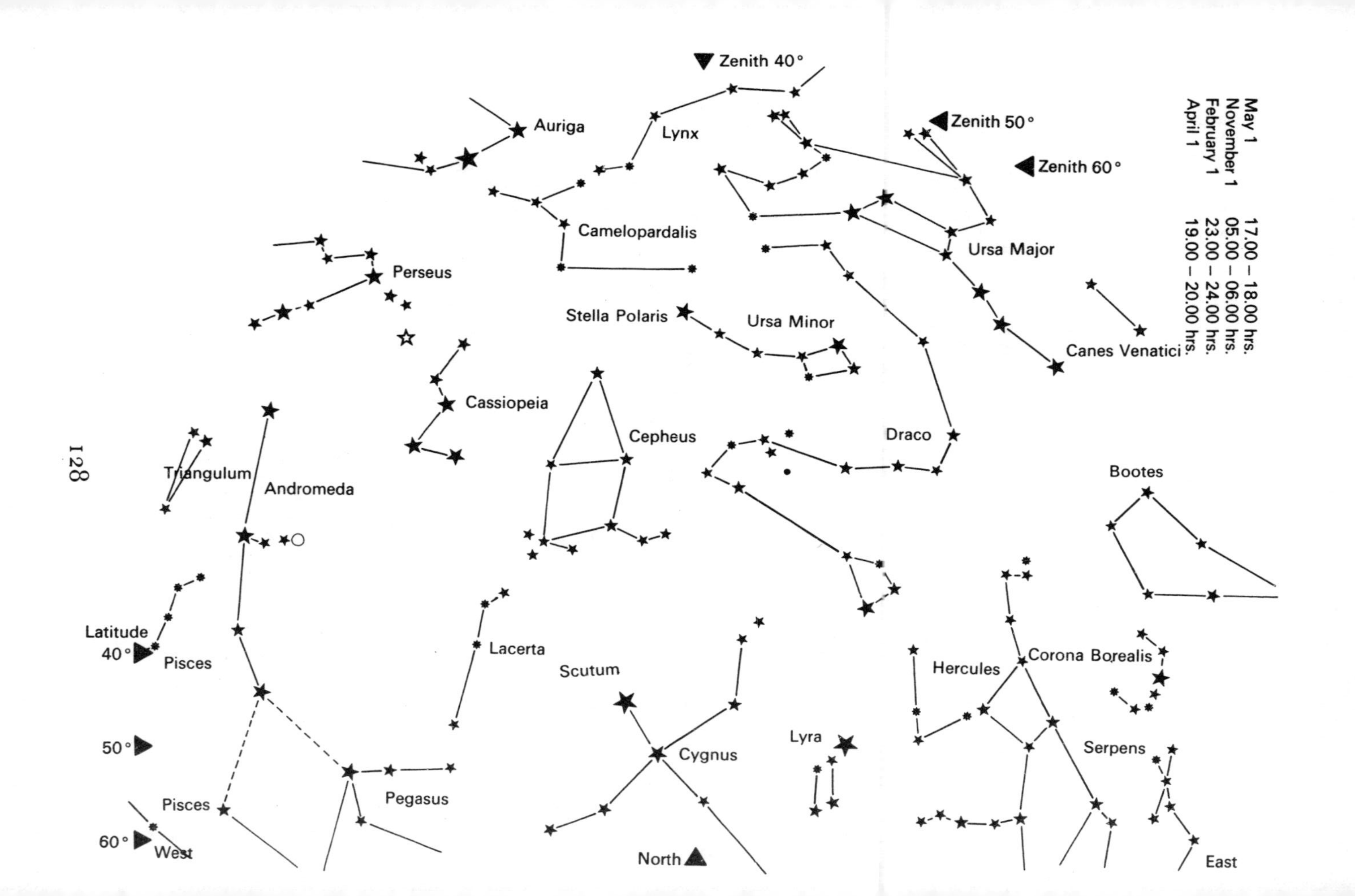

May 1 17.00 – 18.00 hrs.
November 1 05.00 – 06.00 hrs.
February 1 23.00 – 24.00 hrs.
April 1 19.00 – 20.00 hrs.
Zenith 40°
Auriga
Lynx
Zenith 50°
Zenith 60°
Camelopardalis
Ursa Major
Perseus
Stella Polaris
Ursa Minor
Canes Venatici
Cassiopeia
Cepheus
Draco
Triangulum
Andromeda
Bootes
Latitude
40°
Pisces
Lacerta
Scutum
Hercules
Corona Borealis
Lyra
Cygnus
Serpens
50°
Pisces
Pegasus
60°
West
North
East

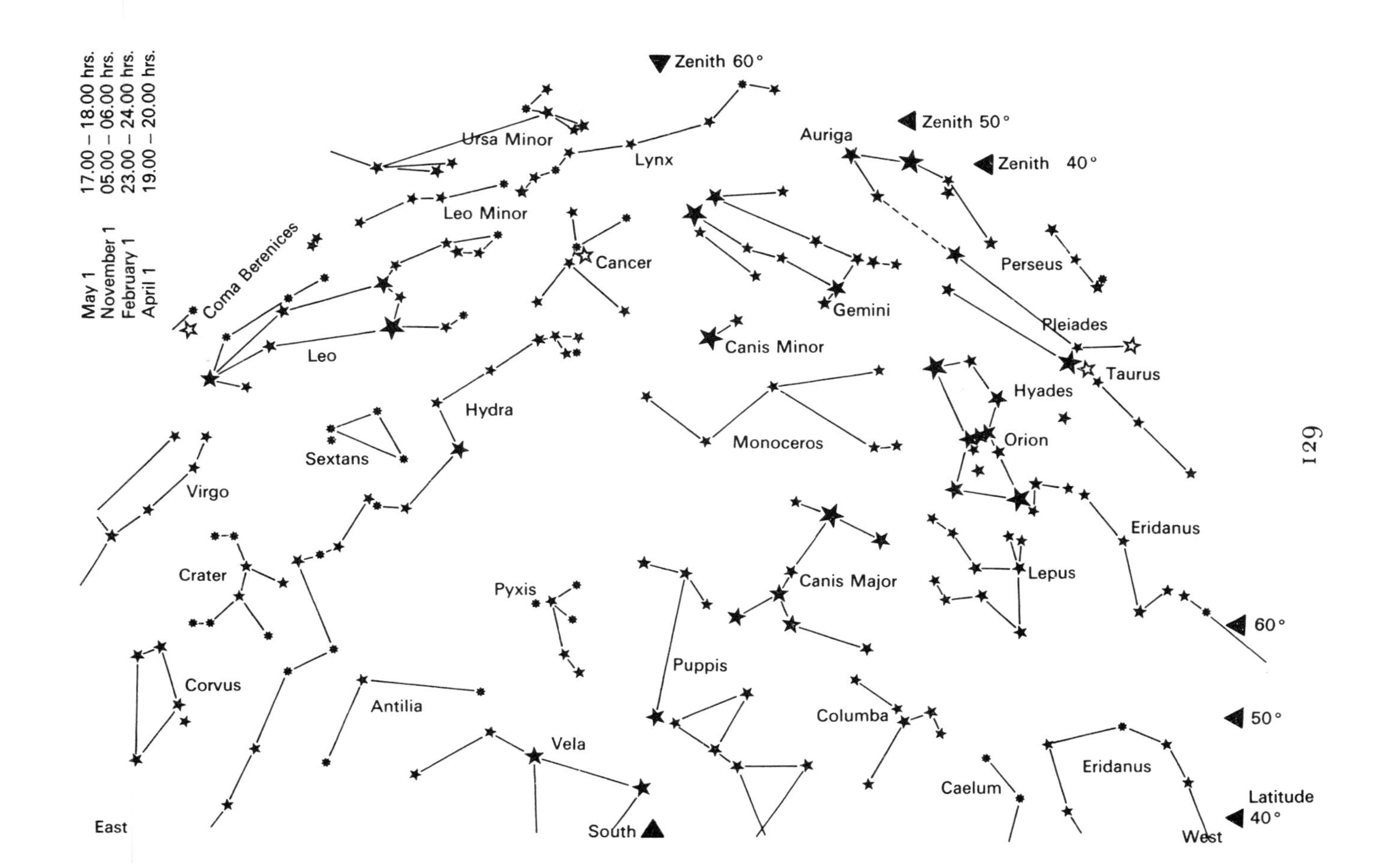
May 1 17.00 – 18.00 hrs.
November 1 05.00 – 06.00 hrs.
February 1 23.00 – 24.00 hrs.
April 1 19.00 – 20.00 hrs.
Zenith 60°
Ursa Minor
Lynx
Auriga
Zenith 50°
Zenith 40°
Leo Minor
Coma Berenices
Cancer
Perseus
Gemini
Leo
Canis Minor
Pleiades
Taurus
Hyades
Hydra
Monoceros
Orion
Sextans
Virgo
Eridanus
Crater
Canis Major
Lepus
Pyxis
60°
Puppis
Corvus
Antilia
Columba
50°
Vela
Eridanus
Caelum
Latitude 40°
East
South
West

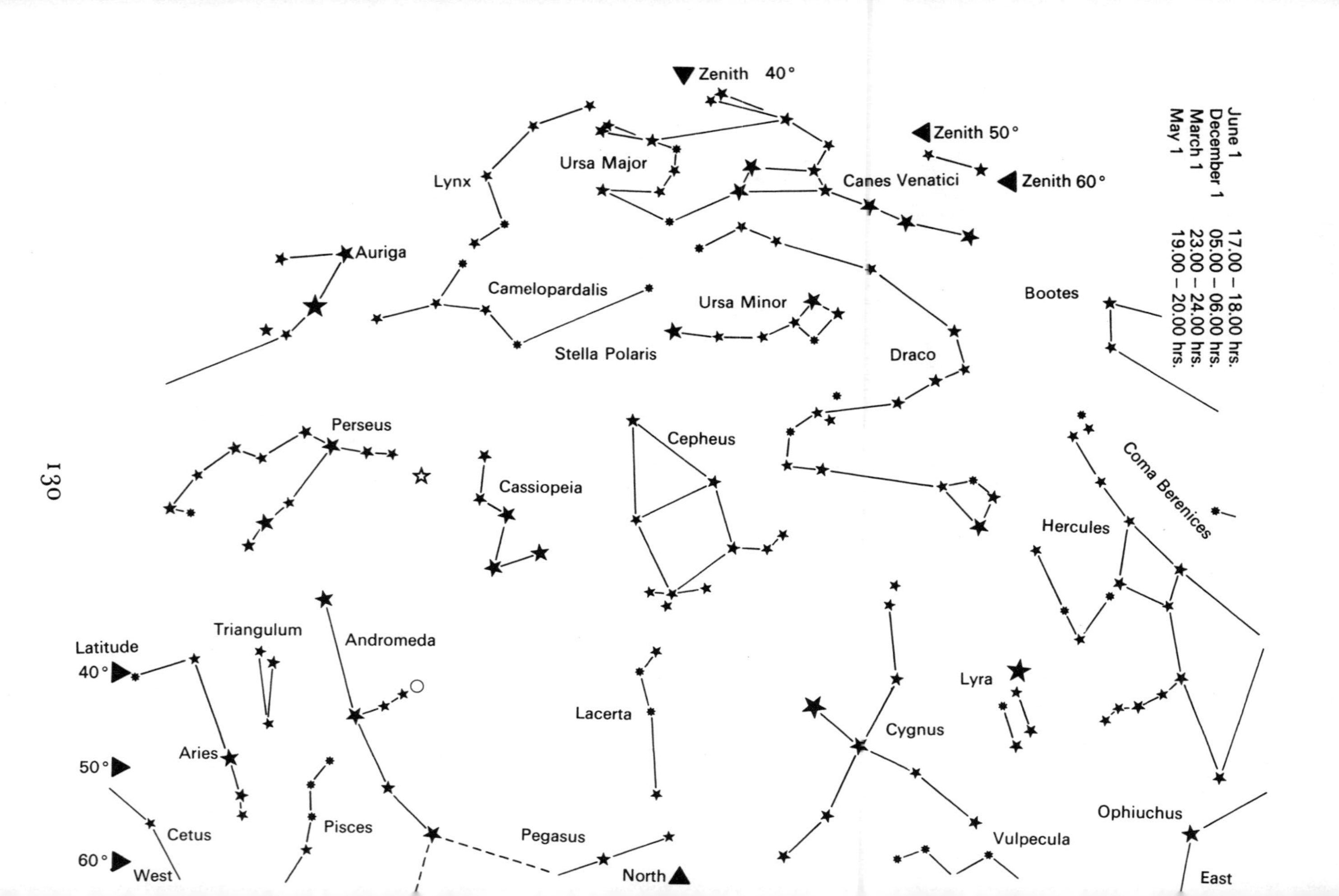
June 1 17.00 – 18.00 hrs.
December 1 05.00 – 06.00 hrs.
March 1 23.00 – 24.00 hrs.
May 1 19.00 – 20.00 hrs.
Zenith 40°
Zenith 50°
Zenith 60°
Ursa Major
Lynx
Canes Venatici
Auriga
Camelopardalis
Ursa Minor
Bootes
Stella Polaris
Draco
Perseus
Cepheus
Cassiopeia
Coma Berenices
Hercules
Latitude
40°
Triangulum
Andromeda
Lacerta
Lyra
Cygnus
Aries
50°
Pisces
Cetus
Pegasus
Ophiuchus
Vulpecula
60°
West
North
East

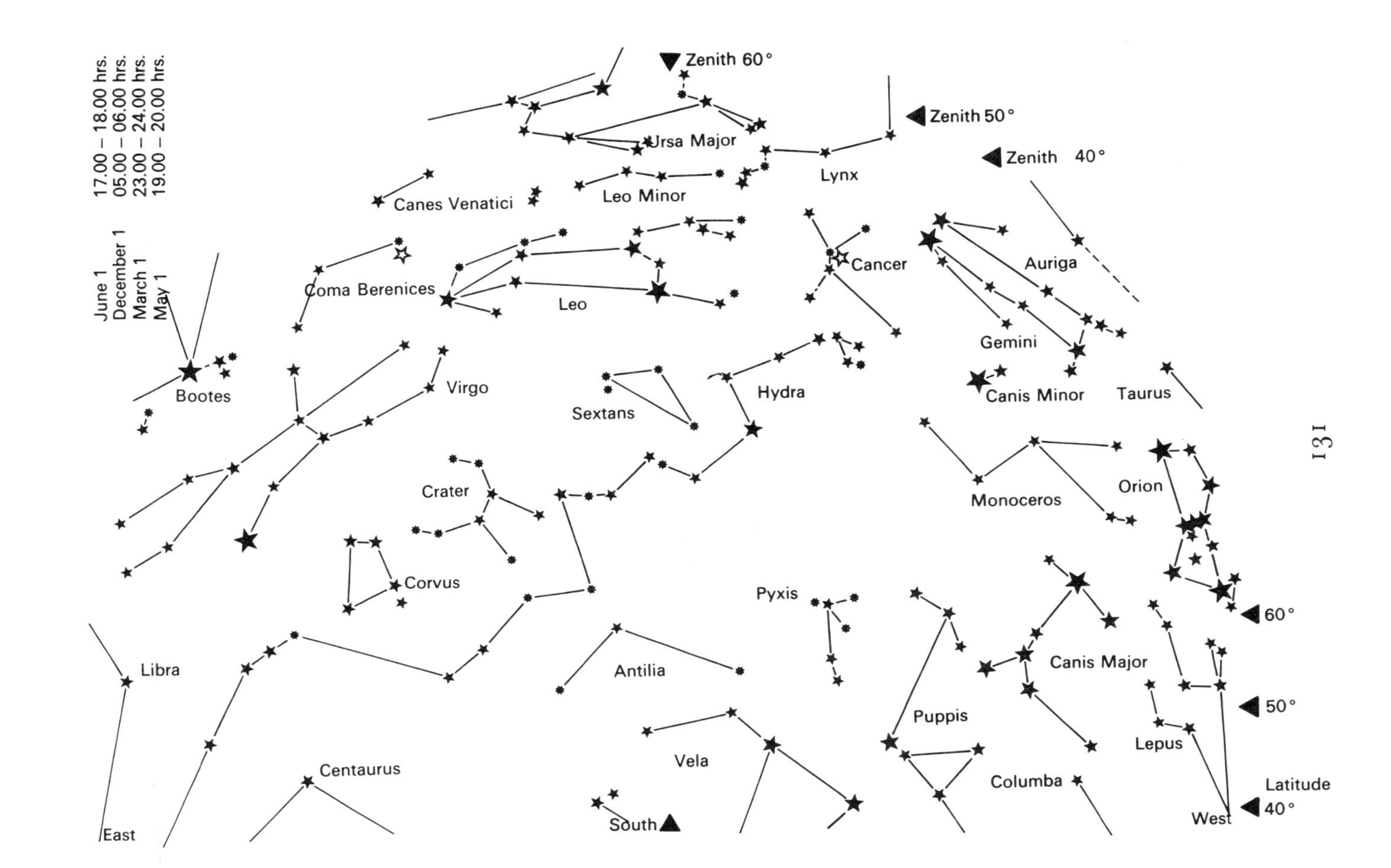

June 1 17.00 – 18.00 hrs.
December 1 05.00 – 06.00 hrs.
March 1 23.00 – 24.00 hrs.
May 1 19.00 – 20.00 hrs.
Zenith 60°
Zenith 50°
Zenith 40°
Ursa Major
Lynx
Canes Venatici
Leo Minor
Cancer
Auriga
Coma Berenices
Leo
Gemini
Bootes
Virgo
Sextans
Hydra
Canis Minor
Taurus
Crater
Monoceros
Orion
Corvus
Pyxis
60°
Libra
Antilia
Canis Major
Puppis
50°
Lepus
Vela
Centaurus
Columba
Latitude
40°
West
East
South

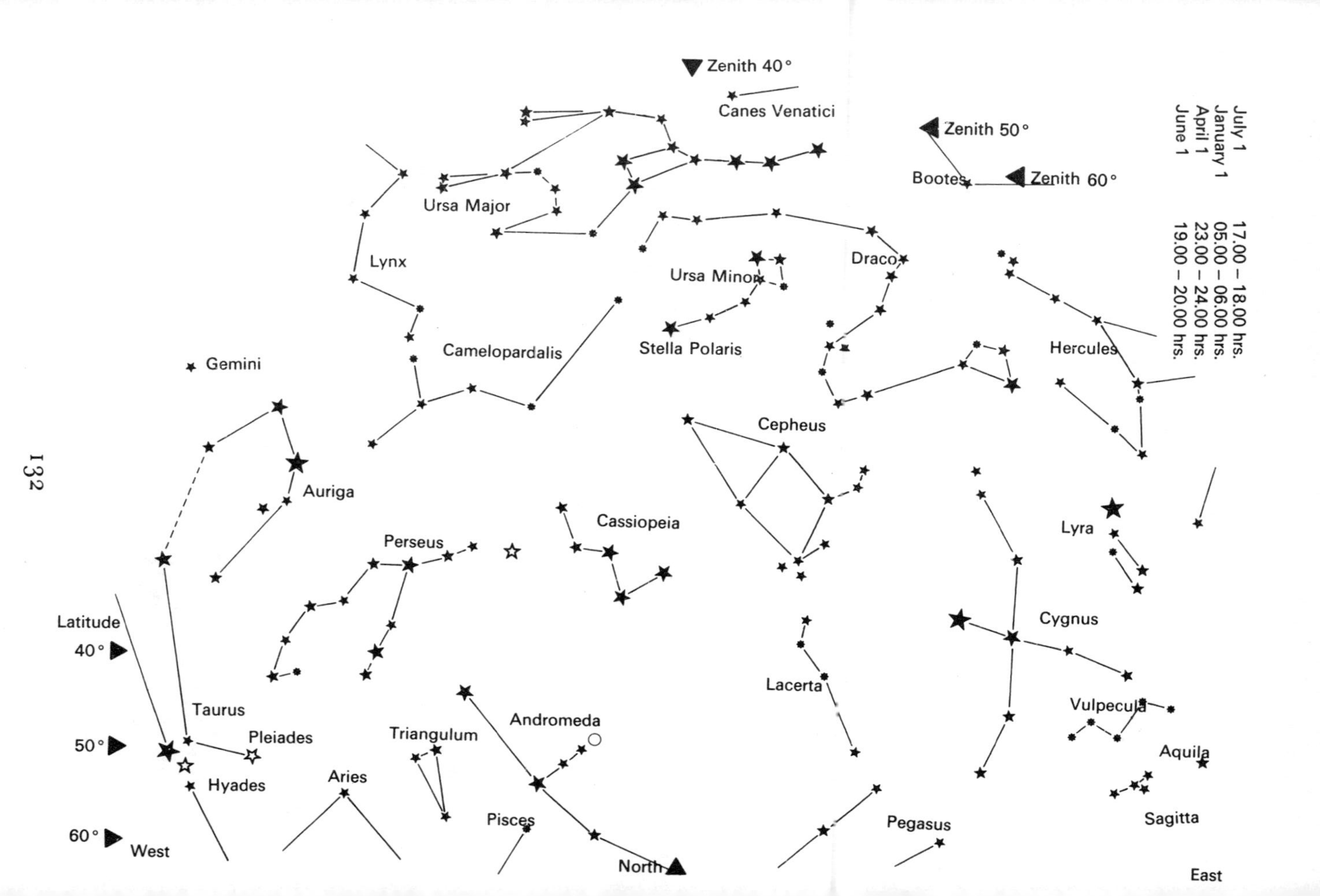

Zenith 40°
Canes Venatici
Zenith 50°
Bootes
Zenith 60°
July 1 17.00 – 18.00 hrs.
January 1 05.00 – 06.00 hrs.
April 1 23.00 – 24.00 hrs.
June 1 19.00 – 20.00 hrs.
Ursa Major
Lynx
Ursa Minor
Draco
Camelopardalis
Stella Polaris
Hercules
Gemini
Cepheus
Auriga
Cassiopeia
Lyra
Perseus
Latitude
40°
Cygnus
Lacerta
Taurus
Andromeda
Vulpecula
50°
Pleiades
Triangulum
Aquila
Hyades
Aries
Sagitta
Pisces
Pegasus
60°
West
North
East

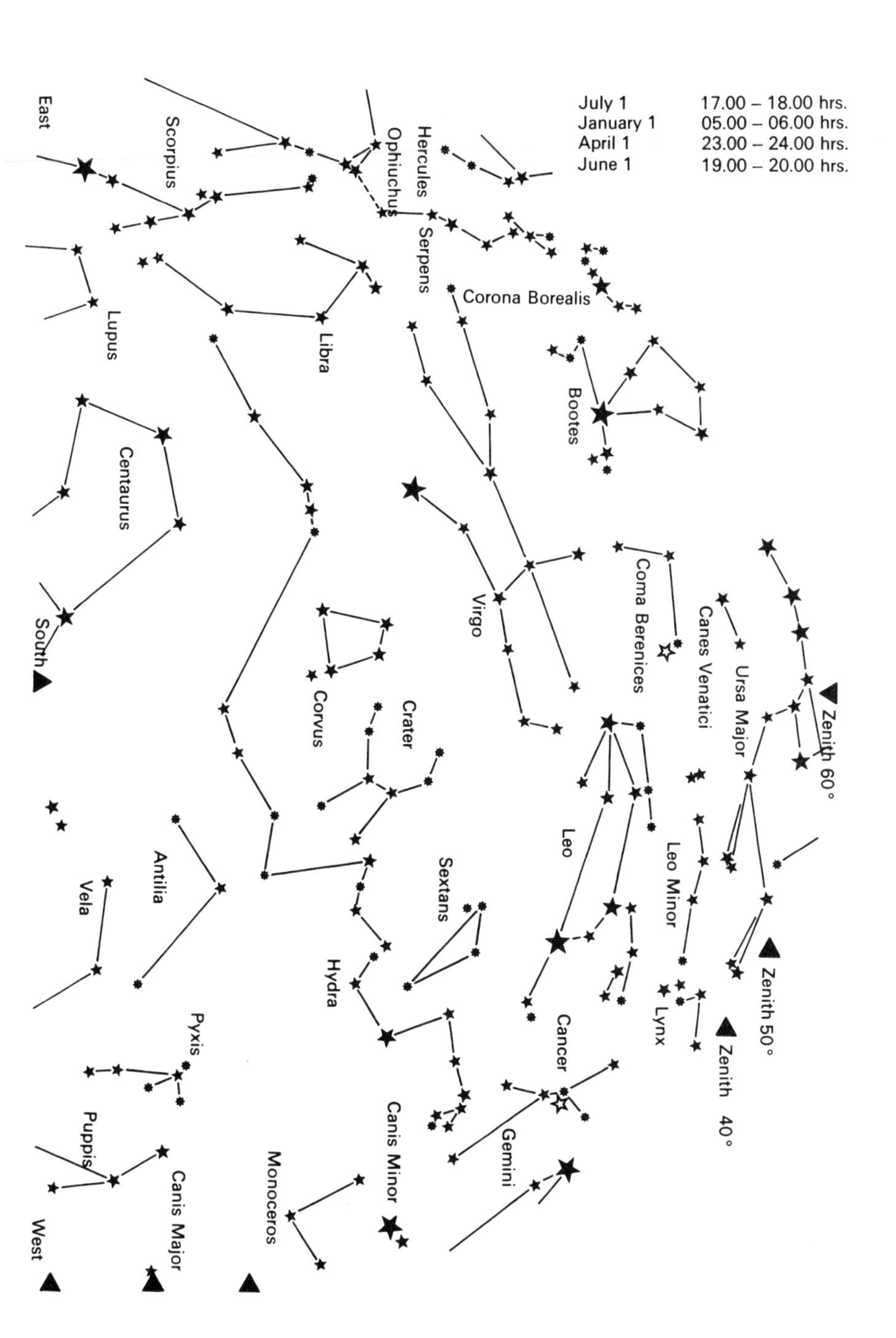
July 1 17.00 – 18.00 hrs.
January 1 05.00 – 06.00 hrs.
April 1 23.00 – 24.00 hrs.
June 1 19.00 – 20.00 hrs.
East
Scorpius
Ophiuchus
Hercules
Serpens
Corona Borealis
Lupus
Libra
Bootes
Centaurus
Virgo
Coma Berenices
Canes Venatici
Ursa Major
Zenith 60°
South
Corvus
Crater
Leo
Leo Minor
Antilia
Vela
Sextans
Hydra
Zenith 50°
Lynx
Cancer
Zenith 40°
Pyxis
Canis Minor
Gemini
Puppis
Canis Major
Monoceros
West

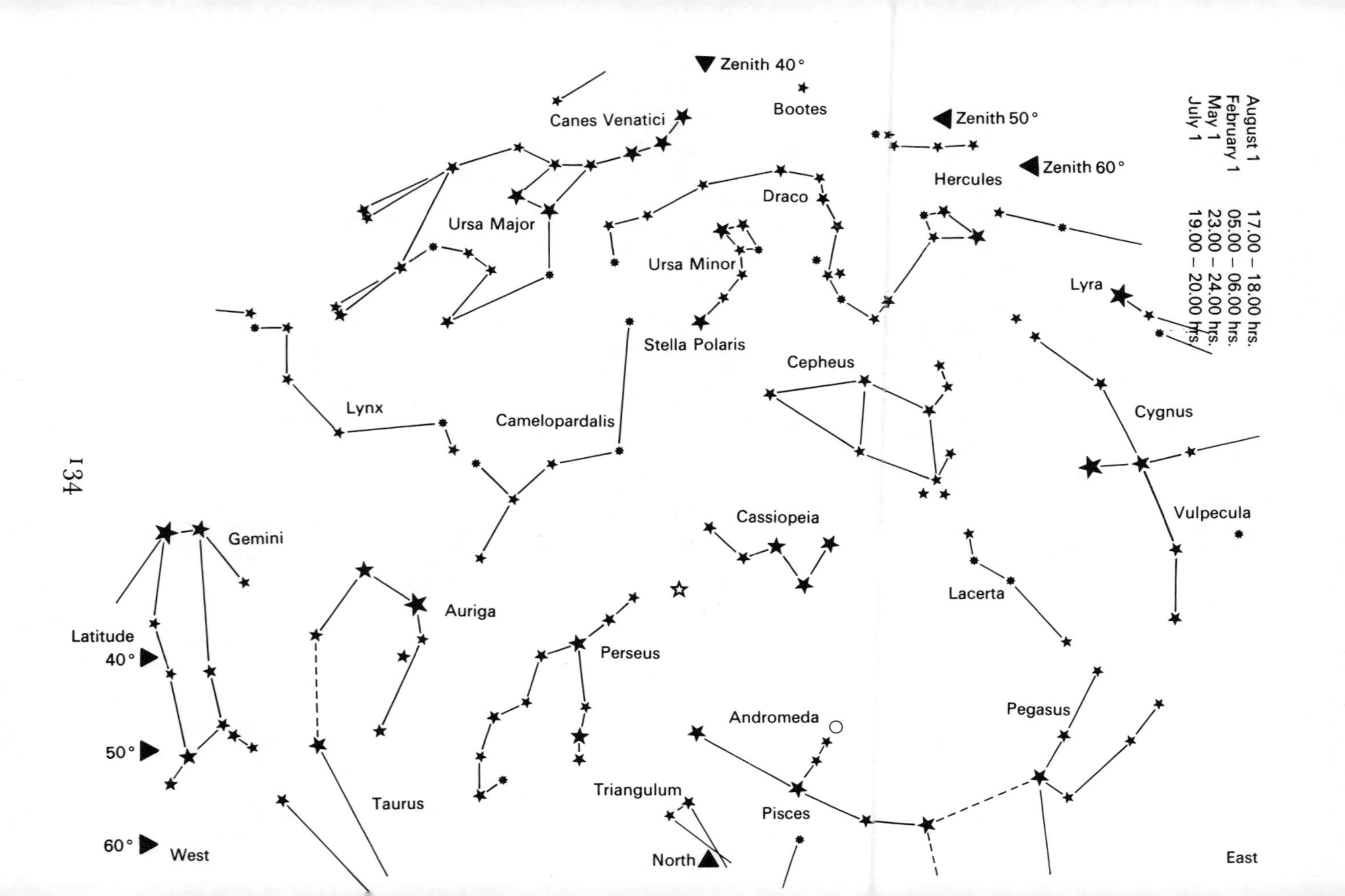
Zenith 40°
Canes Venatici
Bootes
Zenith 50°
Zenith 60°
Hercules
Draco
Ursa Major
Ursa Minor
Lyra
Stella Polaris
Cepheus
Lynx
Camelopardalis
Cygnus
Cassiopeia
Vulpecula
Gemini
Auriga
Lacerta
Latitude
40°
Perseus
Andromeda
Pegasus
50°
Triangulum
Taurus
Pisces
60°
West
North
East
August 1 17.00 – 18.00 hrs.
February 1 05.00 – 06.00 hrs.
May 1 23.00 – 24.00 hrs.
July 1 19.00 – 20.00 hrs.

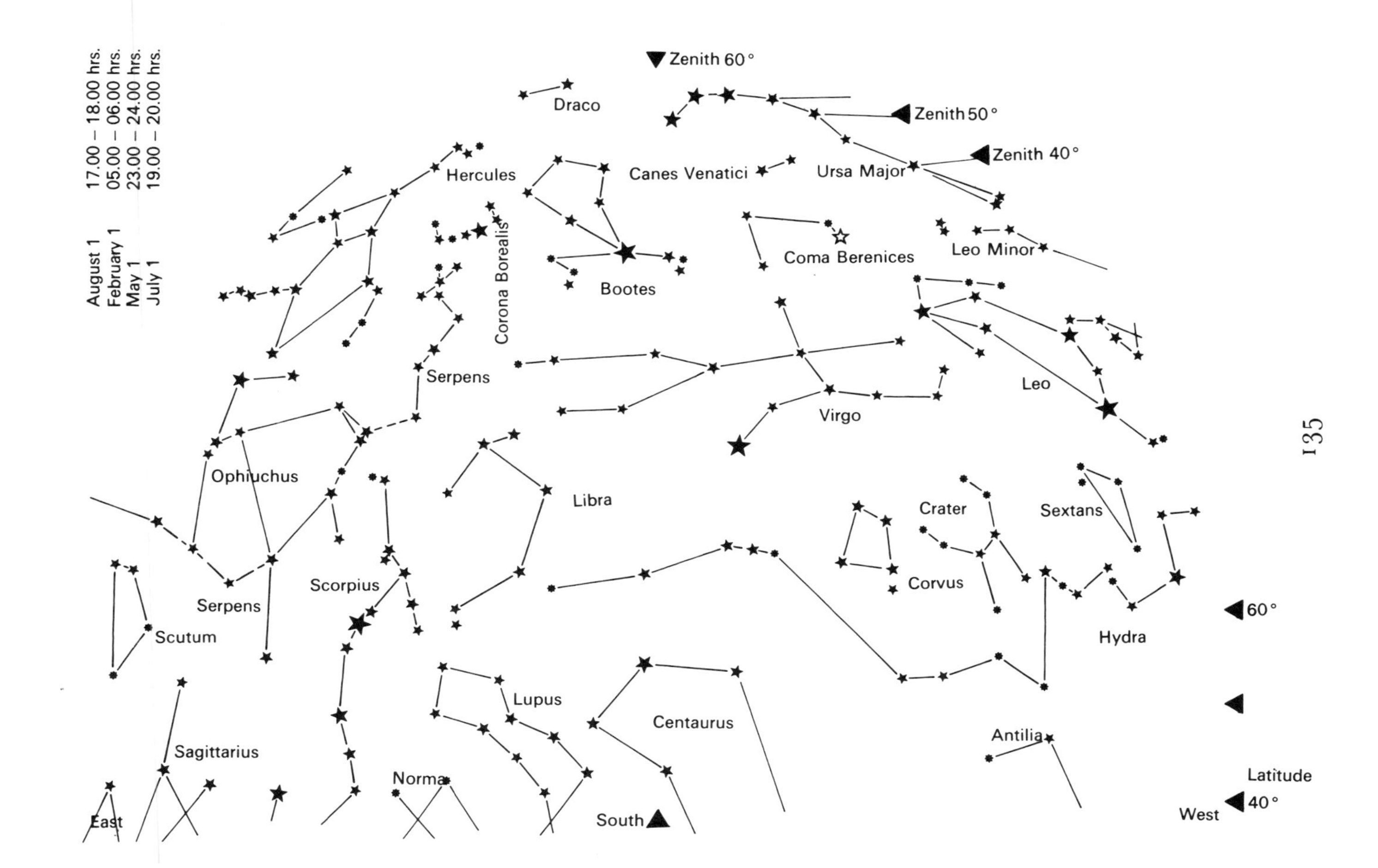
August 1 17.00 – 18.00 hrs.
February 1 05.00 – 06.00 hrs.
May 1 23.00 – 24.00 hrs.
July 1 19.00 – 20.00 hrs.
Zenith 60°
Draco
Zenith 50°
Zenith 40°
Hercules
Canes Venatici
Ursa Major
Corona Borealis
Coma Berenices
Leo Minor
Bootes
Serpens
Leo
Virgo
Ophiuchus
Libra
Crater
Sextans
Scorpius
Corvus
Serpens
60°
Scutum
Hydra
Lupus
Centaurus
Antilia
Sagittarius
Norma
Latitude
40°
West
East
South

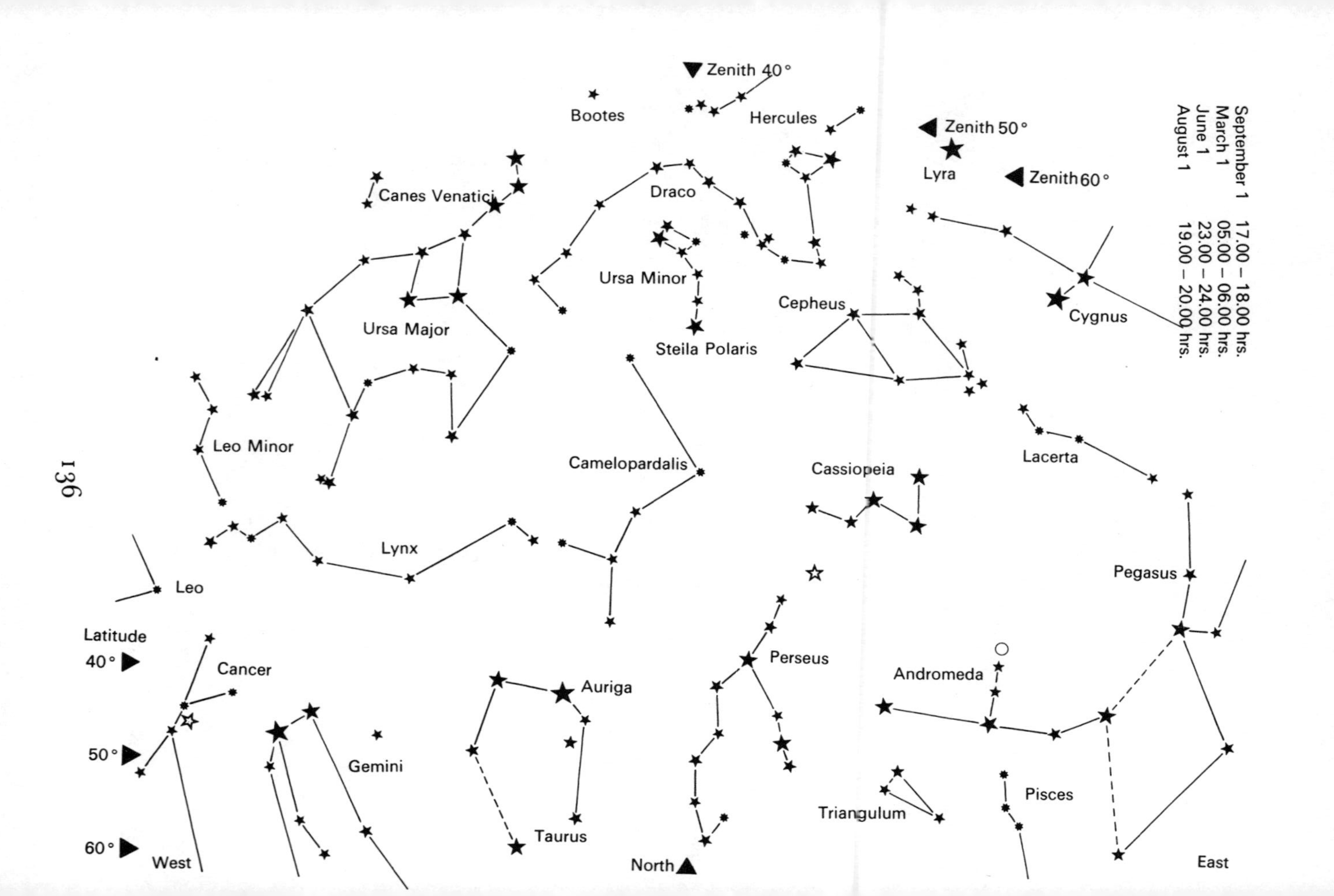

September 1 17.00 – 18.00 hrs.
March 1 05.00 – 06.00 hrs.
June 1 23.00 – 24.00 hrs.
August 1 19.00 – 20.00 hrs.
Zenith 40°
Bootes
Hercules
Zenith 50°
Lyra
Zenith 60°
Canes Venatici
Draco
Ursa Minor
Cepheus
Cygnus
Ursa Major
Steila Polaris
Leo Minor
Camelopardalis
Cassiopeia
Lacerta
Lynx
Pegasus
Leo
Latitude
40°
Cancer
Auriga
Perseus
Andromeda
50°
Gemini
Triangulum
Pisces
60°
West
Taurus
North
East

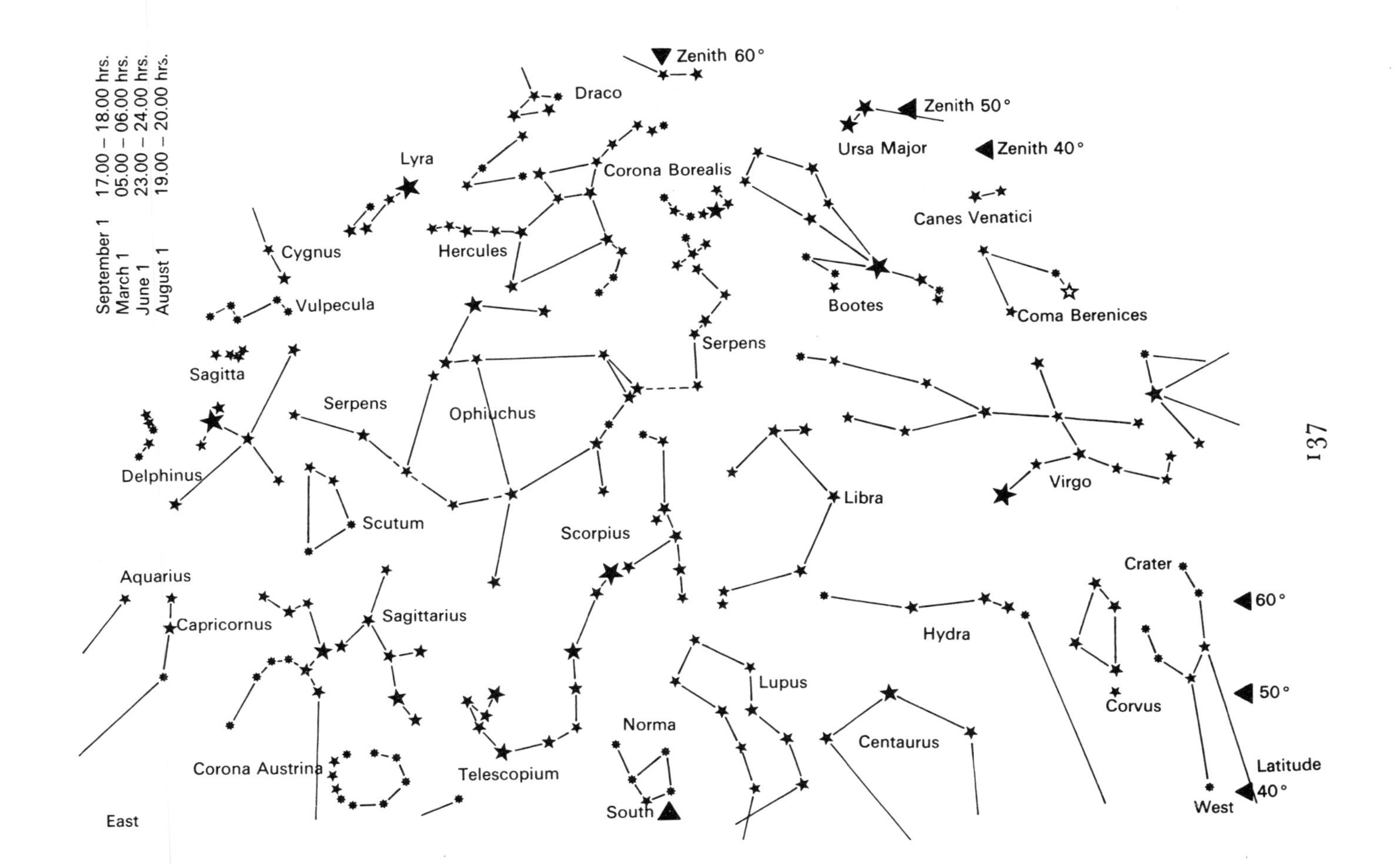
September 1 17.00 – 18.00 hrs.
March 1 05.00 – 06.00 hrs.
June 1 23.00 – 24.00 hrs.
August 1 19.00 – 20.00 hrs.
Zenith 60°
Draco
Zenith 50°
Ursa Major
Zenith 40°
Lyra
Corona Borealis
Canes Venatici
Cygnus
Hercules
Vulpecula
Bootes
Coma Berenices
Serpens
Sagitta
Serpens
Ophiuchus
Delphinus
Virgo
Libra
Scutum
Scorpius
Crater
Aquarius
60°
Capricornus
Sagittarius
Hydra
Lupus
50°
Corvus
Norma
Centaurus
Latitude
Corona Austrina
Telescopium
40°
South
West
East

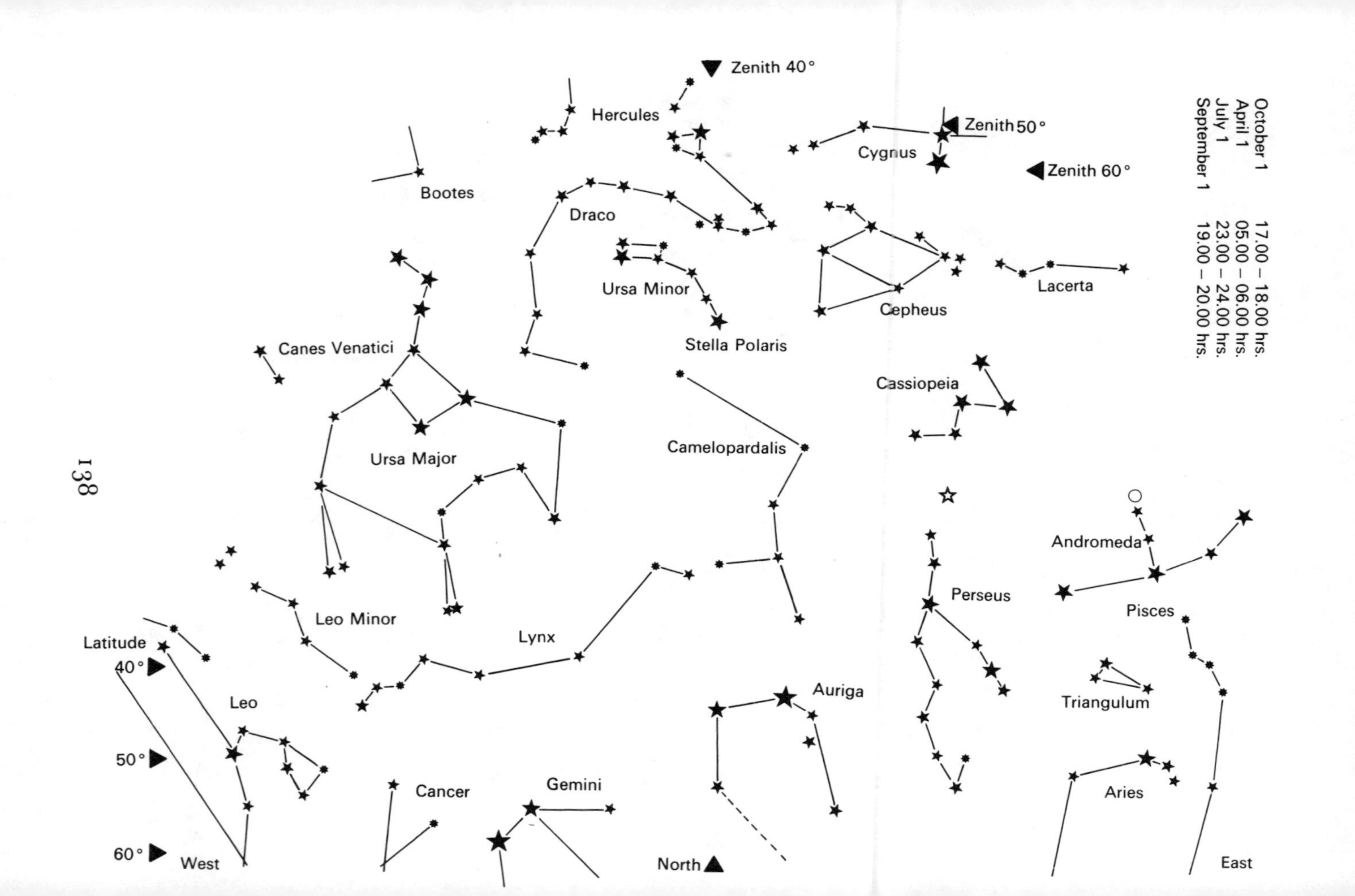
October 1 17.00 – 18.00 hrs.
April 1 05.00 – 06.00 hrs.
July 1 23.00 – 24.00 hrs.
September 1 19.00 – 20.00 hrs.
Zenith 40°
Zenith 50°
Zenith 60°
Hercules
Cygnus
Bootes
Draco
Ursa Minor
Cepheus
Lacerta
Stella Polaris
Canes Venatici
Cassiopeia
Ursa Major
Camelopardalis
Andromeda
Perseus
Pisces
Leo Minor
Lynx
Latitude
40°
Leo
Auriga
Triangulum
50°
Cancer
Gemini
Aries
60°
West
North
East

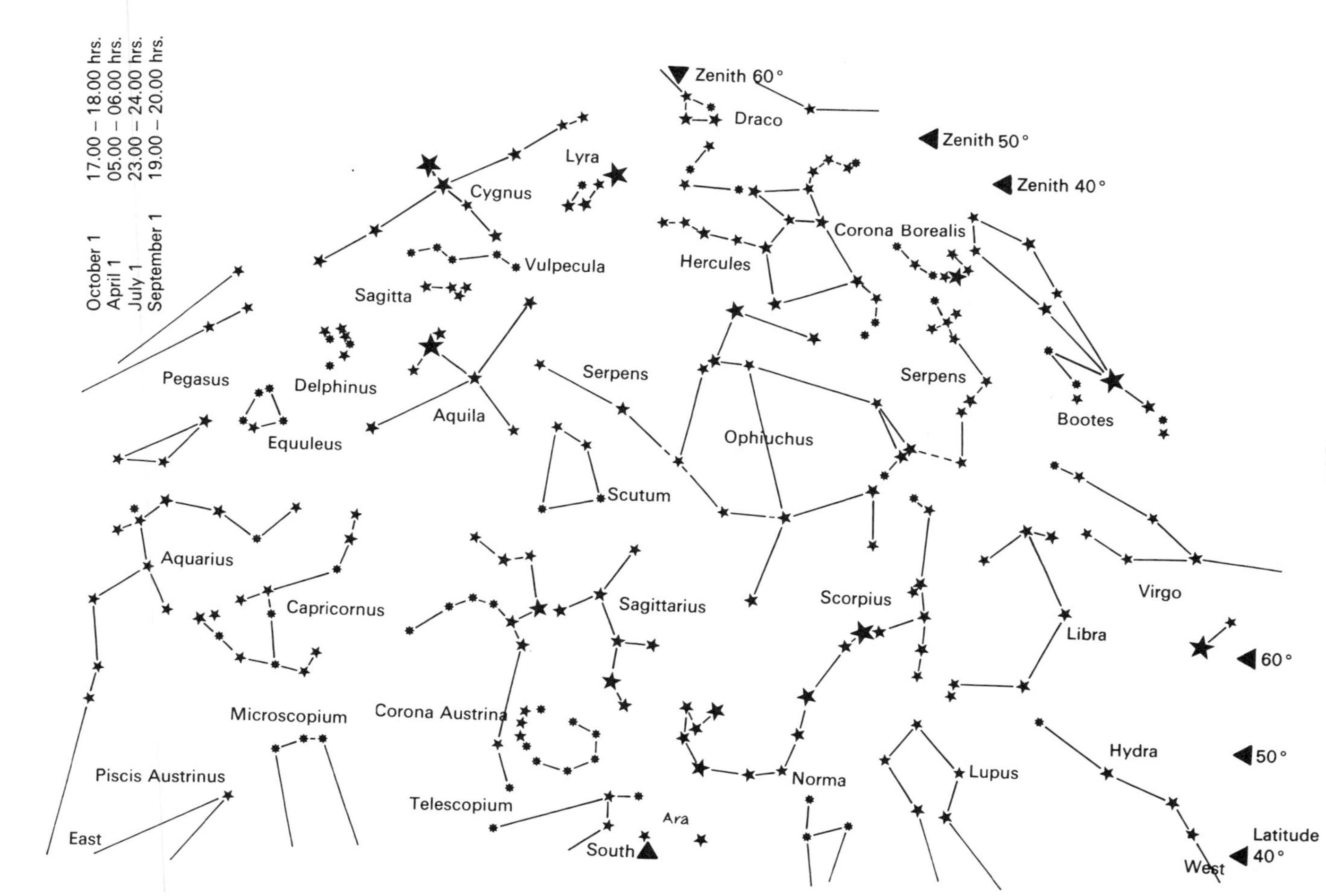
October 1 17.00 – 18.00 hrs.
April 1 05.00 – 06.00 hrs.
July 1 23.00 – 24.00 hrs.
September 1 19.00 – 20.00 hrs.
Zenith 60°
Draco
Zenith 50°
Zenith 40°
Lyra
Cygnus
Corona Borealis
Vulpecula
Hercules
Sagitta
Pegasus
Delphinus
Serpens
Serpens
Aquila
Bootes
Equuleus
Ophiuchus
Scutum
Aquarius
Capricornus
Sagittarius
Scorpius
Virgo
Libra
60°
Microscopium
Corona Austrina
Hydra
50°
Piscis Austrinus
Norma
Lupus
Telescopium
Ara
East
South
Latitude
40°
West

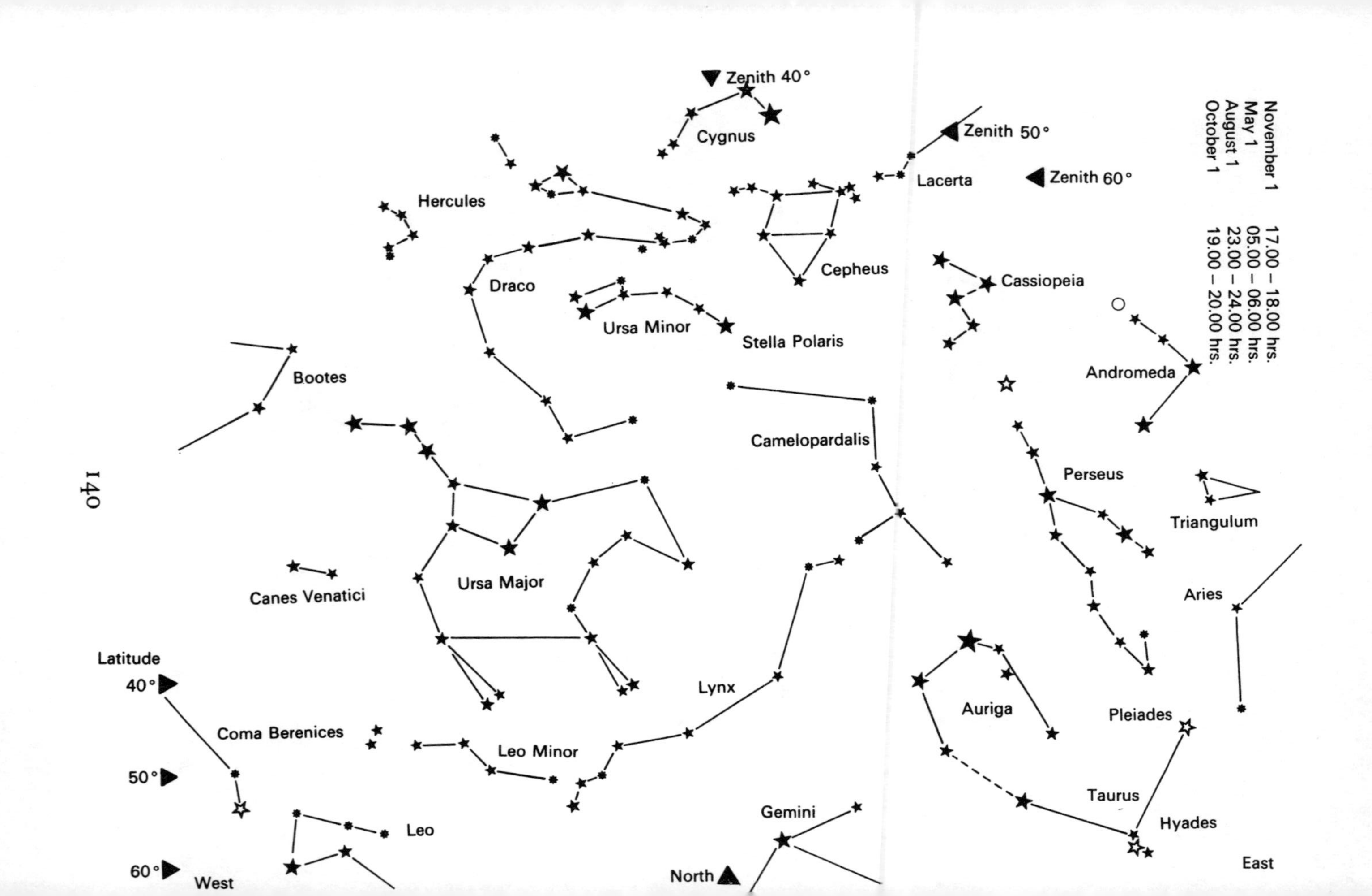

Zenith 40°
Cygnus
Zenith 50°
Zenith 60°
November 1 17.00 – 18.00 hrs.
May 1 05.00 – 06.00 hrs.
August 1 23.00 – 24.00 hrs.
October 1 19.00 – 20.00 hrs.
Hercules
Lacerta
Cepheus
Cassiopeia
Draco
Ursa Minor
Stella Polaris
Andromeda
Bootes
Camelopardalis
Perseus
Triangulum
Ursa Major
Canes Venatici
Aries
Latitude
40°
50°
60°
Lynx
Auriga
Pleiades
Coma Berenices
Leo Minor
Taurus
Hyades
Gemini
Leo
West
North
East

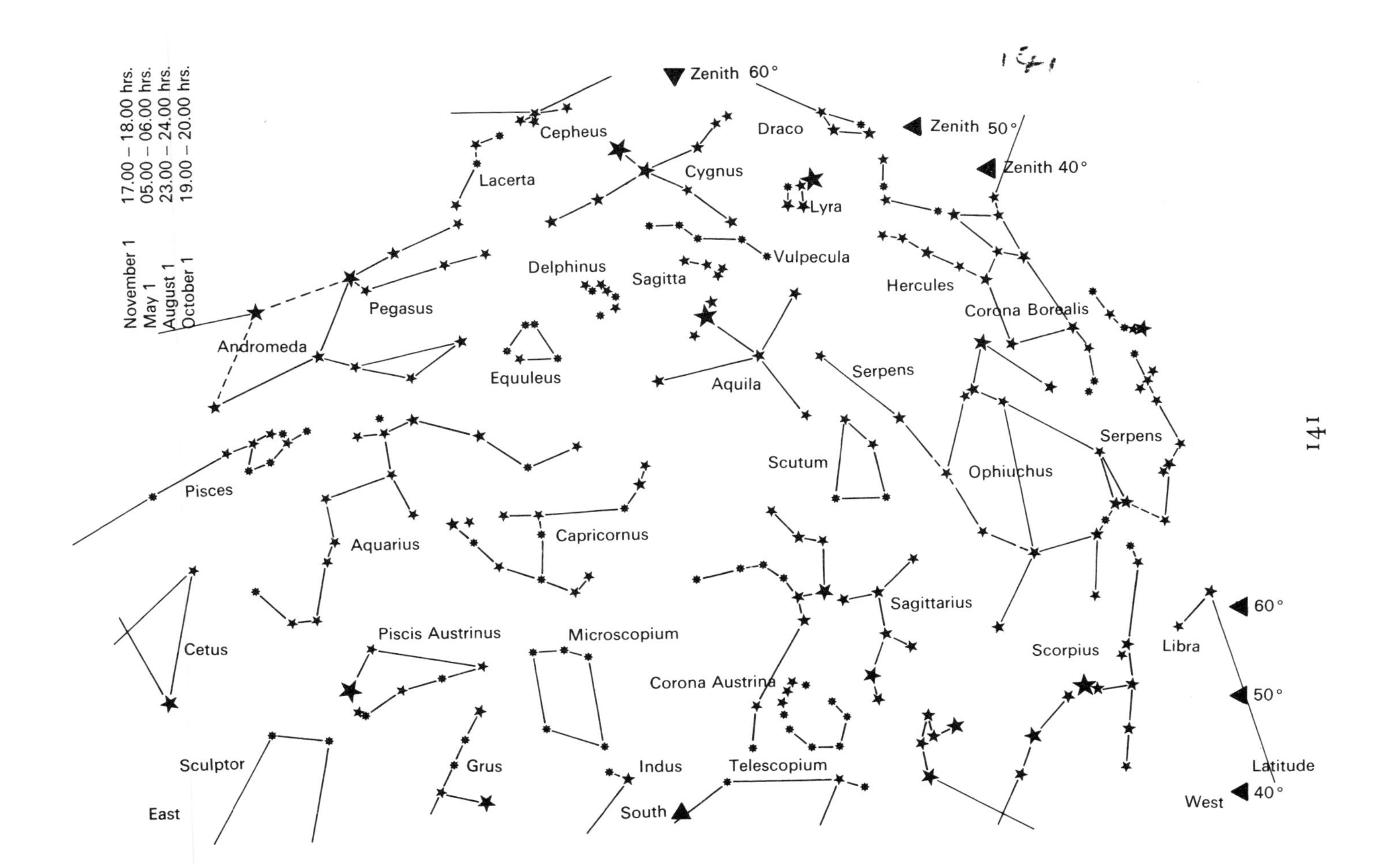

November 1 17.00 – 18.00 hrs.
May 1 05.00 – 06.00 hrs.
August 1 23.00 – 24.00 hrs.
October 1 19.00 – 20.00 hrs.
Zenith 60°
Cepheus
Draco
Zenith 50°
Zenith 40°
Lacerta
Cygnus
Lyra
Vulpecula
Delphinus
Sagitta
Hercules
Pegasus
Corona Borealis
Andromeda
Equuleus
Aquila
Serpens
Serpens
Scutum
Ophiuchus
Pisces
Capricornus
Aquarius
Sagittarius
60°
Cetus
Piscis Austrinus
Microscopium
Scorpius
Libra
Corona Austrina
50°
Sculptor
Grus
Indus
Telescopium
Latitude
East
South
West
40°

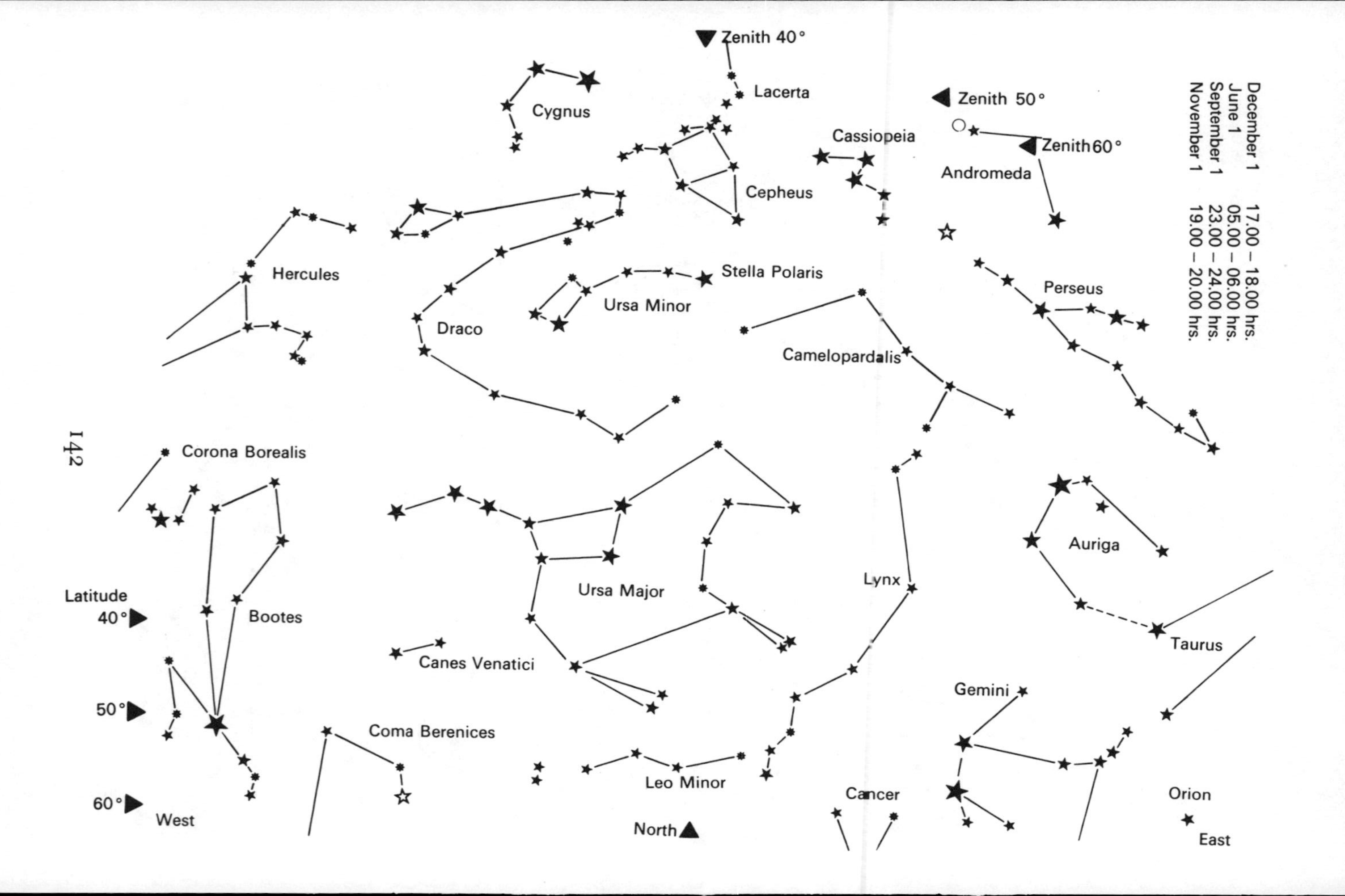

December 1 17.00 – 18.00 hrs.
June 1 05.00 – 06.00 hrs.
September 1 23.00 – 24.00 hrs.
November 1 19.00 – 20.00 hrs.
Zenith 40°
Lacerta
Cygnus
Zenith 50°
Zenith 60°
Cassiopeia
Andromeda
Cepheus
Hercules
Stella Polaris
Perseus
Ursa Minor
Draco
Camelopardalis
Corona Borealis
Auriga
Lynx
Latitude
40°
Bootes
Ursa Major
Taurus
Canes Venatici
50°
Gemini
Coma Berenices
Leo Minor
Cancer
Orion
60°
West
North
East

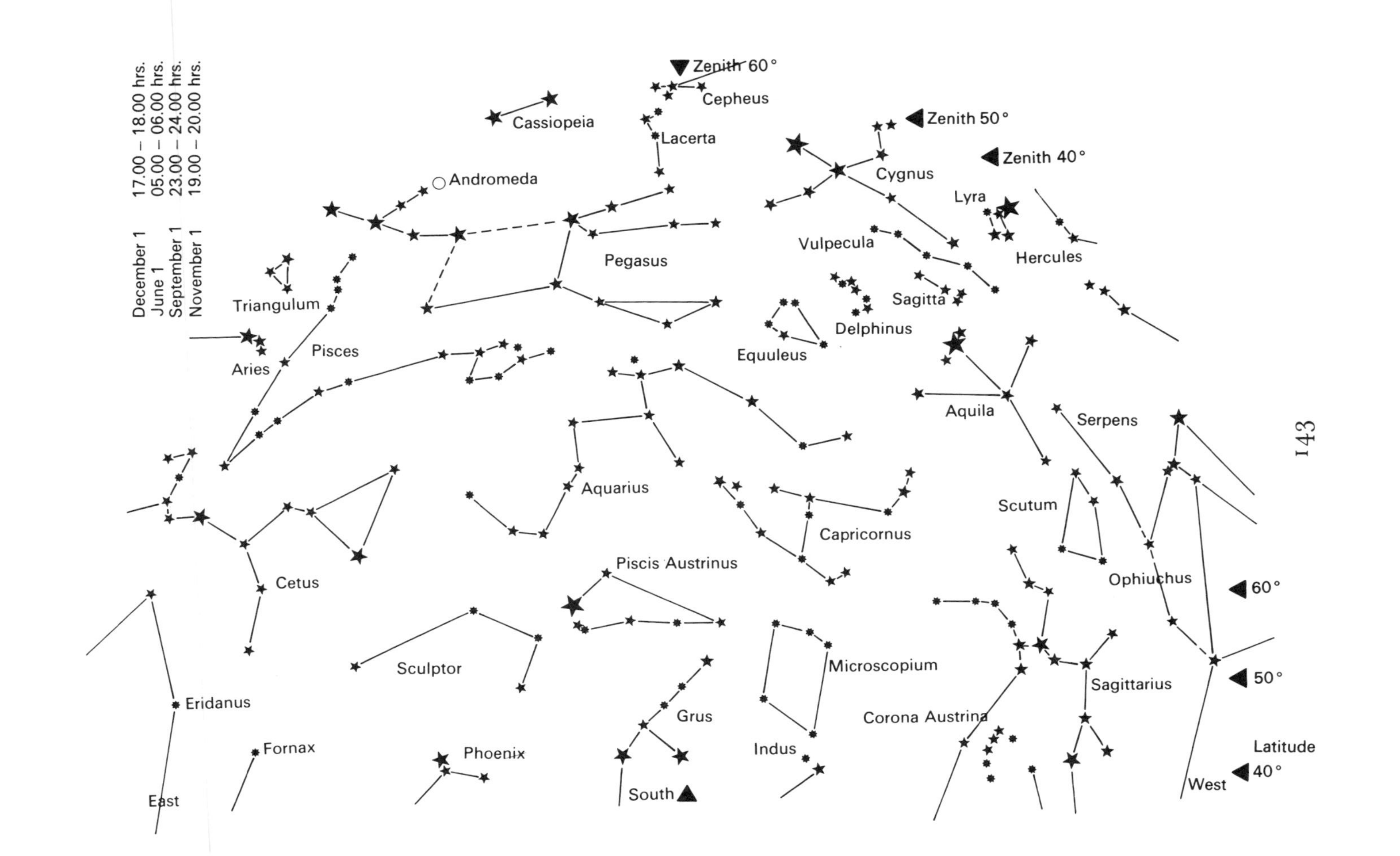
December 1 17.00 – 18.00 hrs.
June 1 05.00 – 06.00 hrs.
September 1 23.00 – 24.00 hrs.
November 1 19.00 – 20.00 hrs.
Zenith 60°
Cepheus
Cassiopeia
Lacerta
Zenith 50°
Zenith 40°
Andromeda
Cygnus
Lyra
Pegasus
Vulpecula
Hercules
Triangulum
Sagitta
Delphinus
Pisces
Equuleus
Aries
Aquila
Serpens
Aquarius
Scutum
Capricornus
Cetus
Piscis Austrinus
Ophiuchus
60°
Sculptor
Microscopium
Sagittarius
50°
Eridanus
Grus
Corona Austrina
Fornax
Phoenix
Indus
Latitude
East
South
West
40°

Index